Managing Indoor Air Quality

Managing Indoor Air Quality

Shirley J. Hansen, Ph.D.

Published by
THE FAIRMONT PRESS, INC.
700 Indian Trail
Lilburn, GA 30247

Library of Congress Cataloging-in-Publication Data

Hansen, Shirley J., 1928 -
Managing indoor air quality / by Shirley J. Hansen.
p. cm.
Includes bibliographical references (p.).
Includes index.
ISBN 0-88173-107-2
1. Indoor air pollution. 2. Air quality management. I. Title.
TD883.1.H36 1991 697.9'3--dc20 90-14049
CIP

Managing Indoor Air Quality / By Shirley J. Hansen.

Published by The Fairmont Press, Inc.
700 Indian Trail
Lilburn, GA 30247

Printed in the United States of America

10 9 8 7 6 5 4 3 2 1

ISBN 0-88173-107-2 FP

ISBN 0-13-553124-1 PH

Distributed by Prentice-Hall, Inc.
A division of Simon & Schuster
Englewood Cliffs, NJ 07632

Prentice-Hall International (UK) Limited, London
Prentice-Hall of Australia Pty. Limited, Sydney
Prentice-Hall Canada Inc., Toronto
Prentice-Hall Hispanoamericana, S.A., Mexico
Prentice-Hall of India Private Limited, New Delhi
Prentice-Hall of Japan, Inc., Tokyo
Simon & Schuster Asia Pte. Ltd., Singapore
Editora Prentice-Hall do Brasil, Ltda., Rio de Janeiro

Contents

PREFACE

Concerns over Indoor Air Quality (IAQ) have driven it to the forefront of work place problems for the 1990s. Health, economic and legal matters associated with IAQ seem destined to make it a dominant problem for developers, owners and managers of commercial and institutional properties well into the next century. Indoor air problems have certainly made the engineer's and facility manager's jobs more difficult and has put the owner at greater risk.

Aside from the obvious humane concerns, the costs associated with achieving indoor air quality seem destined to rise. There is an increasing element of risk and liability associated with poor IAQ. Employers are being plagued by "sick building" associated absenteeism and lower productivity. Employee and occupant lawsuits are on the rise.

The IAQ problem is real and it is not going to go away. The revised ASHRAE standard 62-1989, Ventilation for Acceptable Indoor Air Quality, will continue to force us all to consider the indoor environment even in "healthy" buildings. This book is designed to help those who are responsible for managing that environment. It encompasses the whole range of contaminants, thermal conditions and other factors that contribute to, or detract from, occupant comfort and a productive work place. Focusing on IAQ issues from a management perspective, this book is not intended to be a detailed, technical book. Rather, it is a practitioner's handbook designed to help owners, managers, operators or anyone responsible for operating and maintaining a facility. While not directed primarily at the design professionals, those who offer owners technical and financial support will find it beneficial to examine the IAQ issue from management's point of view.

Managing Indoor Air Quality is structured as a guide as well as a reference document to treat existing indoor air problems effectively, and to help prevent costly IAQ problems from occurring. Finding solutions to IAQ problems is often a complex, multifaceted, multidisciplined endeavor. A single discipline approach from the environmental engineer, the industrial hygienist or the medical doctor, unfortunately tends to narrow the control and treatment options. This book cuts across these professions without the specificity of any one discipline, to offer those concerned with the total facility a broader approach.

For all of the progress we have made in recent years, there are still large gaps in IAQ knowledge. The process, however, is no longer new and experiences in almost every type of facility exist to help us. The book is rich in practical, pragmatic suggestions throughout the procedural guidelines.

The reader is cautioned that every IAQ situation is unique and the procedures discussed herein offer only general guidelines. When problems persist, it is always prudent to call in a special IAQ consultant and/or team to examine a specific situation. Reference herein to any specific commercial product, process, or service by trade name, trademark, manufacturer, or otherwise, does not necessarily constitute or imply its endorsement, recommendation or favoring by the author.

As IAQ concerns have erupted around us, the means to identify, control and treat a myriad of pollutants has mushroomed. Researching, sifting and sorting this burgeoning data was a mammoth task made easier by many people. In particular, I would like to acknowledge the assistance received from Elizabeth Agle and David Mudarri, U.S. Environmental Protection Agency; Michael Crandall, National Institute of Occupational Health & Safety; Patricia Rose, U.S. Department of Energy; Robert Bartley, Delaware Energy Office; Alton Penz, Building Owners Management Association; John Mahoney and Charles Lane of Honeywell; Gary Berlin of Nortec; Jay Santos of Facilities Dynamic Engineering; and Ronald Beckett, Anne Arundel County Public Schools. A very special thanks goes to James Hansen and Hope Worley of Hansen Associates and to F. William Payne of The Fairmont Press.

Shirley J. Hansen

Chapter 1

Indoor Air Quality: A Growing Concern

Through the years, man has built increasingly elaborate boxes to protect himself from the elements. Designed to keep out the rain and snow, warm him in the winter and cool him in the summer, he now lives and works inside these boxes up to 90 percent of the time. Rather than hold environmental hazards at bay, however, he has trapped himself in a chemical soup of contaminants that might make him sick, even kill him.

Buildings don't always protect their occupants from pollution. Just the opposite, the molds, fungi, dust and toxic gases on the inside may well exceed those outdoors. By shielding ourselves from the outside environment, we have created an inside environment with a whole new set of problems.

WHEN BUILDINGS BECOME SICKENING

It usually starts with building occupant's complaints of headaches, nausea, dizziness, sore throats, dry or itchy skin, sinus congestion, nose irritation or excessive fatigue. Suddenly the words that are rapidly becoming the bane of building owners, facility managers and engineers are pronounced: Sick Building Syndrome (SBS). A building is generally defined as sick if 20 percent or more of the building's occupants exhibit such symptoms and the complaints persist for more than two weeks -- particulary, if the symptoms disappear when the sufferers leave the building for the weekend.

The term "syndrome" is used by the medical profession to indicate a number of symptoms occurring together. The exact relationship existing among those symptoms may be unknown. In a sick building, symptoms experienced by occupants are those often associated with the respiratory tract or they may manifest themselves as headaches, dizziness, nausea, lethargy and fatigue. Other symptoms are revealed in areas where the body is directly exposed, such as dry or itchy skin and eye irritation. A sick building can aggravate existing illnesses.

An associated problem is Building Related Illness (BRI). BRI is defined as a building associated, clinically verifiable disease. If signs of actual illness are present and can be attributed to a condition in the facility, the building can be classified as BRI.

The key difference between SBS and BRI is that specific SBS contaminants may not be known. SBS is diagnosed when complaints and symptoms are clearly associated with building occupancy, but no causal agent can be positively identified. Complaints are often resolved by increased ventilation, by more effective controls/substitutions of possible sources and by improved maintenance. In almost every case, BRI is an advanced stage of SBS. The dirt, dust, moisture and stagnant water typical of the poor maintenance that causes SBS is an ideal home for the bacteria or virus looking for an amplification site.

It is highly unlikely that a building will get to BRI stage without first going through SBS.

WHY NOW?

What has made IAQ the hot topic of the 1990s?

With the higher energy costs of the 1970s, new construction and building modifications brought tighter envelopes and increasing reliance on mechanical ventilation. By reducing the intake of outside air, pollutants that were already there have been concentrated and their effects on humans have become more obvious. At the same time, construction materials and furnishings have brought more contaminants into homes and the work place. Technological changes have made copiers, printers, computers, facsimile machines, etc. common place in the office. Operating and cleaning this equipment have brought more pollutants into the office. They have also caused dramatic changes in office procedures prompting ergonomic and organizational stress problems, which have seemed to heighten health problems associated with indoor air pollution.

Against this backdrop, greater press coverage has raised public awareness of indoor contaminants. Furthermore, the federal government has gotten involved. Under the Superfund Bill (P.L. 99-499), Congress authorized the Environmental Protection Agency (EPA) to research radon and indoor air quality. A proliferation of guidelines and regulations are apt to follow. Federal actions are expected to dwarf previous asbestos efforts.

IAQ is costly business. U.S. Army research findings have suggested that IAQ related health problems are costing an estimated $15 billion a year in direct medical costs and about 150 million lost workdays. Garibaldi and Dixon's review of the Army data lead them to conclude that at least $59 billion of indirect costs could be added to the price tag. In 1988, Time reported an out-of-court settlement of close to $600,000 had been made to an occupant in a new office building in Goleta, California, who claimed to have lost consciousness and suffered permanent brain injury due to formaldehyde fumes.

Concerns regarding health, productivity, absenteeism, vacant facilities and the threat of lawsuits are destined to make indoor air quality (IAQ) the dominant issue for building owners and facility operators through the 1990s. It will also remain a critical concern for architects, professional engineers and consultants who offer them technical support.

HUMANE CONCERNS

Since early in this century, episodes of extreme air pollution in highly industrialized and urbanized locations have focused attention on the quality of the air we breathe and associated environmental health concerns. Studies of such episodes and demonstrated increased incidence of respiratory disease and mortality in populations exposed to such pollutants as nitrogen oxide, sulfur oxide and particulates prompted laws to establish specific ambient air quality standards designed to protect the public's health from this type of pollution.

One study of outside air and associated health effects by the Harvard School of Public Health discovered indoor air added another dimension in determining exposure levels. In its Six Cities Study, Harvard researchers found children 6 to 10 years of age living with cigarette-smoking parents had more respiratory illnesses and impairment of lung functions than children living with non-smokers.

The Harvard study and others have shown that indoor environments can play a critical role in exposure levels and health outcomes. Moreover, it has become increasingly evident that the levels of many pollutants may be higher inside buildings than outside.

More Than Just Numbers

Reducing the misery, illness and loss of life to mere statistics risks diminishing the human suffering experienced by individuals due to indoor air pollutants. A few numbers, however, can help document the broad impact air quality has on our lives. In 1987, Platts-Mills estimated 15 percent of the admissions to hospital emergency rooms each year, or 500,000 patients, could be attributed to dust mites and other airborne allergens. The most sensational health problems associated with IAQ, Legionnaire's disease, continues to plague facility managers. Writing in the late 1980s, Fang reported that the Legionella species of bioaerosols, (from cooling towers, hot-water systems and even hot-tubs in residences) account for 8 to 10 percent of community-acquired pneumonia, or 50,000 to 60,000 illnesses, each year. Humidifier fever and hypersensitivity pneumonitis is estimated to occur annually in 1 to 4 percent of the 27 million office workers, running the estimates as high as 1.1 million per year. When health effects attributed to volatile organic compounds (such as formaldehyde), radon, environmental tobacco smoke, etc. are added in, the numbers keep growing. Aside from the human suffering these numbers represent, they suggest economic woes of serious dimensions.

ECONOMIC ISSUES

Corporations with extensive office complexes, such as Xerox, have found that approximately 35 percent of their total operational budget for an office facility is devoted to personnel. Actions, which save few cents in other segments of the facility budget but have a negative effect on personnel are, therefore, apt to have a negative net effect on the bottom line.

Saving Energy and Losing Money

If reduced ventilation saves 10 percent of a 4 percent energy budget, total costs have been cut .4 percent. If that action should cause a 3 percent increase in absenteeism, there would be a net loss in the total operational budget of .65 percent. That's poor economy. There are, of course, ways to optimize energy efficiency and indoor air quality that will be discussed later.

Lost Productivity

We have all observed in ourselves and others that people who are ill do not function as well. They are less productive. While reports of IAQ problems often mention lower productivity, the effects on occupant productivity have not been measured and fully documented; however, it is reasonable to assume that these effects can be very substantial. Isolated studies and logic would suggest that IAQ problems, which have an adverse effect on health, also have a negative impact on productivity.

When office personnel become ill in a given suite of offices and productivity drops, the company moves out. In the greater Washington, D.C. area, a 22 story SBS building stood empty for months while team after team tried to identify the indoor air pollutants and sources causing the health problems. Costs associated with a sick building that stands empty can be staggering. Furthermore, it poses an unending nightmare for the owner, architect, engineer and facility manager. Owners cannot afford empty buildings. Companies cannot afford higher absenteeism and lower productivity. However, legal costs may prove to be the greatest economic burden of all.

LEGAL IMPLICATIONS

In today's litigious society, the specter of legal action always looms large. As a harbinger of things to come, the Wall Street Journal in February, 1988, reported:

> **"SICK BUILDINGS' LEAVE BUILDERS AND OTHERS FACING A WAVE OF LAWSUITS**
>
> More office workers are filing lawsuits, claiming they were made ill by indoor air pollution from such things as insect sprays, cigarette smoke, industrial cleaners, and fumes from new carpeting, furniture, draperies and copiers.

Most journals for design professionals, (architects, engineers and interior designers) have run articles by attorneys urging them to limit their IAQ liabilities.

As early as June, 1986, The Professional Liability Perspective, published by a prominent Northern California insurance broker, warned design professionals, "The potential liability problems posed by indoor pollution are compounded by the fact that the pollution exclusion in your policy of professional liability insurance is all-compassing. It extends to every form of environmental contamination imaginable. The risk is simply not insurable." After offering several protective measures design professionals might take, the article suggested that the problem primarily rests with the owner. The broker's newsletter urged architects and engineers to recommend to the owner that he/she retain a qualified air quality consultant.

The article concluded, "Indoor pollution is seen by many close to the profession as something of a new frontier for the underemployed at the plaintiff's bar. Your challenge is to take steps to remove yourself from the path of the almost inevitable stampede. You can do so if you heed and act upon the early signs of danger on the horizon." If the advent of newsletters such as Indoor Air Pollutant Law Report, can be viewed as a harbinger of things to come and further evidence of "danger on the horizon;" surely, the insurance broker's warning should extend beyond design professionals to building owners and operators.

INDOOR AIR POLLUTANTS

What's at the root of the problem? What causes buildings to become "sick?" Indoor air contaminants, which affect the health of occupants, can be divided into particles (solids or liquid droplets) and gases or vapors. While tobacco smoke immediately comes to mind as an indoor air pollutant, other contaminants that have been treated extensively in the press, such as asbestos fibers, radon and formaldehyde are readily recognized as contributors to the problem. Allergens, such as pollen and fungi, are well known to asthmatics.

Carbon dioxide (CO_2) is sometimes listed as a pollutant, but does not present a problem unless the contamination level is very high. CO_2 does, however, serve as an excellent surrogate in testing for other gases.

The list is long. Cigarette smoke alone is known to contain 4,700 chemical compounds, including several that have been shown to be highly toxic in animal tests, and 43 suspected carcinogenic compounds. The chapter on classifying IAQ problems discusses the major indoor pollutants. Appendix A addresses each of these pollutants separately including a description, probable sources, health effects, existing national laws, regulations, standards or codes. As appropriate, notification, possible control methods and monitoring are discussed by contaminant.

Some published papers on IAQ differentiate between contaminants and pollutants; citing a contaminant as anything foreign in the air or water and a pollutant as something with an adverse effect on humans. At this point, it seems to be a forced distinction. Webster uses pollutant as a synonym for contaminant and the terms are used interchangeably in this book.

TREATING THE PROBLEM

Nearly every article on indoor air quality mentions the energy efficient "tight" building. Since returning to those good ol' leaky buildings does not seem likely, the knee jerk reaction has been to recommend an increase in ventilation. Ventilation is a key ingredient, but those who offer ventilation as the only means of treating these contaminants fail to consider the problem in sufficient depth. An August 1988 quote from a trade journal article is rather typical of the cursory remedies put forth to resolve a complex problem:

> Practically speaking, attributing the problem to an unidentified contaminant or contaminants rather than saying, simply that the ventilation is inadequate doesn't make much difference if the prescription is to deal with ventilation...

But knowing what the contaminants are does make a difference! Several studies have shown that increased air changes per hour (ACH) have little or no correlation to some pollutants, such as some volatile organic compounds. In other instances, as with radon, increased ACH can actually increase contaminant levels. Figure 1-1 shows ACH plotted against radon concentrations for 250 buildings taken from three studies.

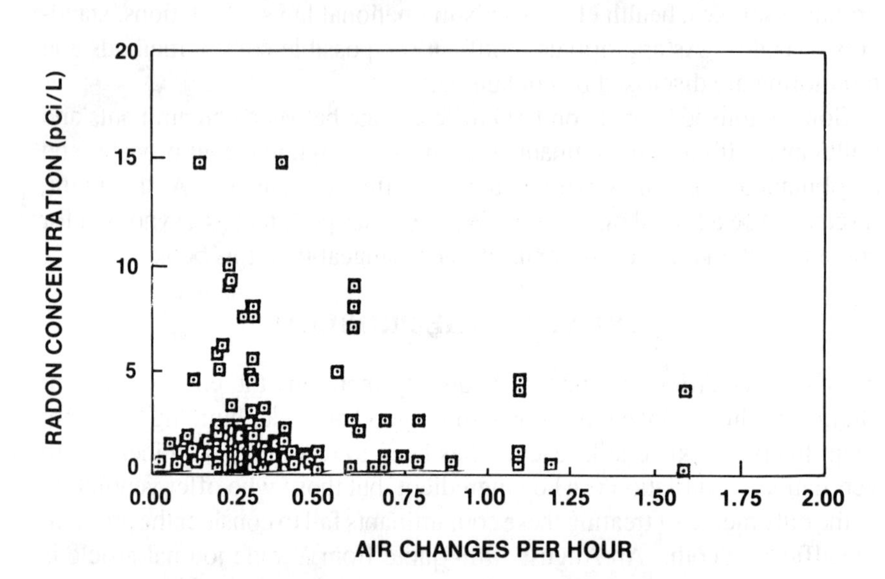

Source: New York State Energy Office based on the work of LBL, W.S. Fleming, Wagner

Figure 1-1. Air Changes Vs. Radon Concentrations

IAQ CONTROLS

Understanding the ways pollutants can be controlled puts ventilation as a solution in perspective and enlarges upon the maintenance opportunities.

There are six ways to control indoor air pollution:

1. Removal or substitution at the source.
2. Encapsulation, or otherwise interfering with the materials' ability to give off pollutants;
3. Filtration and purification of contaminants;
4. Time of use of a possible contaminant;
5. Education and training of building occupants, especially operations and maintenance personnel; and

6. Dilution (ventilation) of the indoor air with outside air or filtered recycled air.

The EPA in its 1989 report to Congress stated that source control is the most direct and dependable control option, and the only effective one when strong pollutant sources are present. Source control usually demands some type of investigative procedures.

INVESTIGATION HEADACHES

The greatest "headaches" associated with indoor air problems may come from investigating suspected indoor air quality concerns. Facility managers, directors of maintenance or operations, and plant engineers, trying to diagnose a problem are apt to get the very headaches they are working to help occupants avoid. Assessing and controlling indoor air pollutants is a complex process. IAQ investigations tend to be complicated by the passage of time, emotional overtones and the fact that symptoms are not easily attributed to a single cause. Furthermore, investigations frequently require a multidisciplinary approach calling, for example, for both medical and engineering expertise.

The National Institute of Occupational Safety and Health (NIOSH), which has the broadest experience in IAQ investigations, has observed that "the application of standard industrial hygiene, medical and epidemiological techniques may prove to be inconclusive." Drawing from its experience in over 500 investigations, NIOSH has developed a solution-oriented approach that progressively eliminates the most frequent causes. The NIOSH approach and other investigative procedures are discussed in Chapter 5, Investigating Indoor Air Problems.

NIOSH's logical approach can guide the layman's first steps as well. Be practical: if it smells like there's a skunk under the house, chances are there's a skunk under the house. Establishing procedures to remove the most likely causes first just might avoid much of the headache and cost associated with diagnosing the problem. The concept carries over to preventive procedures, especially in the area of maintenance, for the easiest remedy of all is: Don't let the skunk under the house in the first place.

ENVIRONMENTAL MEDICINE

Once the investigation has revealed some probable causes, the problems related to IAQ are not over. As with any new field, new labels of expertise are donned by those practicing untested treatment.

Burge, an allergist with the University of Michigan Medical Center, has counseled against the ready acceptance of "clinical ecologists" or "specialists in environmental medicine." Burge observes, "Unfortunately, the methods and theories of these practitioners have not been subjected to rigorous scientific study... The dangers involved in using unproved methods of diagnosis are several." After noting, the patient may not receive the help they actually need or the treatment itself might be deleterious, she concludes, "Finally, the drastic changes in life-style often recommended by environmental medicine practitioners to implement avoidance can be expensive, both financially and emotionally, and probably rarely, if ever, provide a permanent solution to the underlying complaint."

Burge's warning has merit. Every new field has its share of charlatans. These words of caution, however, should not be taken to suggest that any individual or firm self-labeled as clinical ecologists or specialists in environmental medicine are necessarily engaged in some level of quackery. It does remind owners, managers and technical support personnel that the credentials and references of anyone purporting to be an IAQ specialist should be checked. Furthermore, the potential employer of such specialists needs to know enough about indoor air pollutants to know the special expertise required to do the job.

THE MANAGEMENT CHALLENGE

IAQ problems are increasingly prevalent, pervasive and even pernicious. New IAQ standards by ASHRAE will force us all to consider the indoor environment even in "healthy" buildings. Managing that internal environment has become a major administrative responsibility. Nearly all the IAQ literature is written by researchers and consultants for other researchers and consultants. This book has been developed to provide practical guidance to owners, facility managers and those with facility-related responsibilities. It also offers some special insights regarding management concerns for those who provide technical, legal and financial support to practitioners.

Chapter 2

Indoor Air Quality In Retrospect

Students in a New Jersey high school are sent home ill. Workers in a University of Florida Veterinary Teaching Hospital complain of headaches, rashes and a "metallic" taste in their mouths. Office personnel from the Environmental Protection Agency appear at Congressional air quality hearings in silent protest of their environmental work conditions. An assistant director of a New York agency points to his "dried fruit" basket where fruit and vegetables have shriveled into hard, brown contorted shapes in the desert-like office conditions. All these incidents illustrate a problem in common: the quality of indoor air.

Problems with indoor air quality are not new. They undoubtedly reach back to combustion contamination from fires in the caves. A retrospective examination of the many factors leading to our current indoor air quality concerns helps to establish the background for the difficulties posed by internal environments today.

IAQ IN ANTIQUITY

Modern social attitudes toward indoor air quality date back to antiquity. Egyptian historians have found notations of respiratory disease caused by silicate dust from stone cutting as early as 1500 B.C. A thousand years later, Hippocrates described the effects of air on health in his writings. Pliny the Elder, in early Roman times, urged stone masons and asbestos miners to wear masks while working.

The Roman baths were designed to control the volume of water, light, air heat and sound in any given room through room orientation, window placement and the positioning of fountains, plumbing and furnaces. This marriage of the building envelope and its orientation to the heating, ventilation and water distribution systems was a precursor of today's architectural and engineering attempts to provide acceptable indoor air quality.

While concerns regarding comfort and climate did not cease during the early middle ages, they got a fresh impetus late in the period from increased urbanization and the associated pollution problems. Records show that the unbearable stench of the Fleet River caused a Carmelite monastery to petition King Edward I in 1520, for relief from the filth that was interfering with divine services and causing death among the monks.

Large scale energy related indoor air problems can be traced to the burning of coal in the 13th century. By 1273, a law was passed prohibiting the use of coal in London due to the smoke. Parliament petitioned the King to stop the burning of coal altogether in 1306, as the fumes were considered dangerous to health.

Perhaps the most notable incident of this period was the proclamation by King Edward III forbidding the dumping of garbage into the Thames and Fleet rivers when the stench imperiled all London in 1357. Designated as the "Year of the Great Stink," a warm summer brought eutrophic conditions to the rivers, causing excessive algae growth, bacterial depletion of oxygen, and tons of dead fish.

BODY ODOR AND TOBACCO SMOKE: FACTORS OF TRANSITION

Through the years, body odor and tobacco smoke were the prime factors in assessing the quality of the indoor air. Cleanliness has not always been next to Godliness. The decline in the practice of bathing during the Middle Ages, which persisted through the Renaissance, can be attributed to the attitude of the Christians toward cleanliness. The Church attacked the preoccupation with body comfort and attractiveness offered by bathing as tending toward sin. The "odor of sanctity" prevailed and lice were called "pearls of God." St. Paul is said to have observed, "The purity of the body and its garments means the impurity of the soul."

More frequently the lack of personal hygiene during the Renaissance was an economic concern. Poor people worked and slept in the only clothing they owned. While the rich owned more changes of clothes, there is little evidence that they were laundered between wearings.

Smoking tobacco has alternately been accepted and rejected by society and the law. King James I was the first to denounce the habit as a "corruption of health and manners." During the 17th century, most of Europe severely penalized or forbade the consumption of tobacco. As recently as 1911, 14 states prohibited cigarettes for moral and/or health reasons. Today these concerns are reflected in the prohibitions governing the sale of tobacco to minors, and the increasing restrictions on areas where smoking is permitted.

These two indoor air quality factors, body odor and tobacco smoke, strongly influence our standards today. The need for fresh air has historically been measured in the need to counteract human generated pollutants; thus, the common ventilation requirements are for so many cubic feet of air per minute per occupant. The sources of many contaminants today; e.g., building materials, combustion, cleaning supplies, etc. have very little relationship to the number of occupants in a building.

As we move into more modern times, it becomes clear that indoor air quality has had its own "Back to the Future" scenario. With the development of the first nuclear powered submarine in 1954, submarines suddenly had the capability to remain submerged for weeks (or months) at a time. This required a means of controlling, cleaning and revitalizing the quality of indoor air. Through the use of special ventilation and filter systems, air conditioning, chilled water systems, main oxygen supply, CO_2 removal, CO-H_2 burners and atmosphere analyzing systems, the internal air in a modern submarine can be maintained at a designated quality level. The designers of the Nautilus' environmental system were ahead of today's building designers by nearly half a century.

IAQ TODAY

What has brought about the more recent concern for the quality of air we breathe? Virtually every article on indoor air quality points to the energy efficient building as the culprit. Moves to cut utility costs certainly contributed to the problem, but it is much greater, much broader than that. Only in examining the complexity of the problem, can we begin to see the complexity of the solution. Those who attribute the problem solely to the energy efficient building, think increasing ventilation will totally solve the problem. When in fact, increased ventilation in some instances, will not help the problem and could even make matters worse.

The Energy Response

When energy prices climbed by more than 600 percent in the 1970s, schools, with their rigidly constrained budgets, were among the first to seek band-aid approaches to their dilemma. Those "band-aids" frequently showed up as plastic patches over air intake grills and windows.

Schools were not alone in such responses. Little was known about energy efficiency, but the obvious high cost of conditioning outside air made the reduction in air intake a natural choice —right after adjusting the thermostats. Outside air requirements generally dropped from 15 or more cubic feet per minute (cfm)/occupant to 5 cfm/person.

As we became more sophisticated in energy efficient practices, delayed start-ups and early shut downs to let the systems "coast" became standard operating procedures. Ventilation was designed to rely more heavily on mechanical systems with recycled air. The cracks and leaks were patched and insulation was added to walls and ceilings. The cost-saving, energy efficient building, was borne. Unfortunately, we are now beginning to appreciate what a high price tag those cost cutting efforts may carry.

The response to high energy prices carried another cruel twist. As owners sought ways to pay the utility bill, they looked for other places in the facility budget to cut costs. The most elastic part of the budget seemed to be operations and maintenance staff and materials. Those cuts only exacerbated the indoor air problem, for one of the greatest sources of indoor air pollutants is inappropriate or insufficient maintenance. The effects of maintenance cuts were compounded by the energy efficient building, which is far less forgiving of poor maintenance.

Technological Changes

But that wasn't all. Changes in product technology introduced new building materials, products and furnishings into the indoor environment that emitted large numbers of chemicals into the air. Among the culprits were carpet backing and adhesives, wall coverings, paints, stains, paneling and ceiling tile. Occupants exposed to such chemicals often complained of irritation, discomfort and other flu-like symptoms. The effects of many of chemicals emitted from the products are still not fully understood, but many are known or suspected human irritants and some are suspected human carcinogens. Many years may yet go by before the dangers associated with the use of certain chemicals are realized and brought to public attention. Consider, for example, the past and present uses of tobacco, formaldehyde or solvents used in glues.

To this chemical soup, we've added a host of cleaning materials. Even such familiar things as floor wax and the emissions from dry-cleaned clothes can increase contaminant levels.

From the old manual typewriter and hectograph gel, we have progressed to computers and photocopiers. Facsimile machines have become standard office equipment. If all these technological marvels don't emit harmful contaminants in their operation, there is a good chance the cleaning agents used to service and maintain them will. Frequently, this equipment has been added to old facilities that do not have sufficient air flow to satisfy the newer operational needs. It is not unusual to find copiers stuck off in a corner, hemmed in with files, or set in some unventilated office niche. Because the original HVAC design cannot take into account these activities, irritating or dangerous levels of pollutants can accumulate.

Building Use Changes

Changes in building use during the life of a building are common. Altered programs, functions, personnel needs, etc. frequently have necessitated facility modifications. Other changes have been mandated by federal laws, such as handicap access legislation, or through state statutes and local ordinances. New equipment or alterations to the electrical or lighting systems have prompted building changes. Spatial modifications, particularly partitioning the 1970's "open space" facilities, have changed interior layouts. Most of the time, corresponding changes in the mechanical or distribution systems have not kept pace. Most every facility manager is aware of pockets of "stale" air, or areas where over (or under) heating and cooling have resulted from such changes.

More general construction practices have also left their "heritage." For instance, the placement of buildings in radon affected areas without consideration as to how radon migrates into the facility has fostered a whole radon testing and mitigation industry.

HVAC as a Source

The building's own HVAC system may be the culprit. Air pollutants may be moved by the distribution system from an area of the building used for a specialized purpose, such as industrial shops or laboratories, into other areas, such as offices. Polluted outside air, no longer referred to as "fresh air," may be deliberately brought in from loading docks, garages, or picked up from the building's own exhaust. Polluted air, such as automobile exhaust in a parking garage, may find its way up elevator shafts and stairwells into office spaces.

Humidifiers, dehumidifiers, air conditioners, cooling towers and duct linings may be the source of biological contaminants spread by the ventilation system.

The People Factor

Occupants in this changing internal environment have experienced physical, emotional and mental reactions. Frequently, the cause-effect relationships are not fully documented. Synergistic relationships between contaminants and ergonomic or organizational stress are not fully understood. Those examining IAQ related problems, have pointed to such subtle things as the occupants' inability to control their environment. Whether it truly helped or not, there was something innately satisfying about throwing open the window. Job stress and job satisfaction have been suspected as contributors to indoor air complaints; however, Hedge's research in the United Kingdom found no significant relationship. Others have suggested job stress, dissatisfaction or even problems with glare lighting or uncomfortable furniture may have prompted increased use of alcohol or drugs. These substances may constitute intervening variables in any cause-and-effect studies related in indoor air quality.

In 1988, Goldhaber, Polen and Hiatt reported a strong correlation between prolonged exposure (over 20 hours per week) to visual display terminals (VDTs) and miscarriages. The editors of Indoor Air Quality Update, when reporting the study, questioned if the quality of the air may have been a factor. They suggested emissions from the VDTs themselves may result in chemical exposure; or, they postulated, the immediate environment where VDTs are used may vary from other space in the office.

Several studies have observed that women seem to be more susceptible to some indoor air pollutants and are more apt to complain than men. The quality of indoor air in a given facility is not uniform. Clerical workers, who are predominantly women, are often crowded together and they are frequently in interior air supply zones. The perimeter offices generally receive more air supply and more outside air than interior areas. The interior zones are more likely to stay near the warmer end of the design and thermal operating range. Airborne chemicals from the clerical equipment, such as copiers, and materials, such as carbonless copy paper, increase the clerical staff's exposure levels.

The potential relationship of all these factors help to illustrate the problems researchers have in trying to establish causal relationships; and investigators have in trying to pin down certain pollutants as factors in building related illnesses. Indoor air quality research is reminiscent of the admonition, "If you think you have a solution, you probably don't understand the problem."

The Changing Scene

Federal actions to date, especially with reference to asbestos and radon, have been directed primarily at schools and homes. A movement toward the commercial sector can be anticipated. Commercial facility owners have limited concerns related to radon, but asbestos removal from buildings constructed prior to 1972 could turn budgets inside out.

This retrospect also serves to remind us: Amenities are not created equal. Nor should they be. Johnson and Curley list the lobby, elevator, heating system and the bathrooms as, "The basic things that are most important to us as 'spec' office builders." These are emphasized more than other facets because of their importance to prospective tenants.

Focus is gradually shifting from buildings that are self-conscious art forms that may attract tenants to the comfort that will keep them. Mikulina has observed, "We must help builders and developers understand that, while an architect can attract tenants through aesthetics, only a good comfort system will retain tenants through increased productivity."

Recognizing "comfort" as the thermal conditions and air quality, Mikulina concludes, "Insufficient budgets are often allocated to comfort because budgeting decision makers--architects and owners-- don't understand the damage that can be done to a building's financial value by indiscriminately cutting a comfort system's budget. They don't understand because we have failed to communicate this important message."

Air quality has been a concern for 4,500 years. Today, the specter of lawsuits, lost productivity and frequent tenant turnover has garnered more and more attention from owners, facility managers, architects and engineers. It is becoming increasingly clear that it is bad business not "to communicate the important message" that a comfortable healthy environment is a productive environment.

Chapter 3

Management Procedures

The way a complaint is handled can be the most crucial aspect of managing indoor air quality and warrants special attention prior to addressing broader management considerations. Whether the complaint is received by custodial staff, middle management, the CEO or someone talking to a board member at the company picnic, how that initial contact is treated can be critical.

COMPLAINT RESPONSE

Each IAQ episode exacts a price, human and economic. That price almost always far outweighs the costs of establishing an effective complaint response procedure.

A complaint that is lightly dismissed creates an emotional climate that is hard to overcome. A seemingly innocuous complaint that does not receive careful attention can grow into job dissatisfaction and lost productivity. Almost overnight, it can spread into a disgruntled work force, even a union dispute.

A complaint may actually be prompted by a cold or the flu. Or, a newspaper article, coupled with someone's active imagination and encouragement from the coffee lounge brigade, may start things going. On the other hand, the symptoms may be from a very real building-associated illness. In either case, it can't be ignored.

When employees believe they are being forced to work in an area that is injurious to their health, they resent it. The Honeywell IAQ Diagnostics group has found psychological factors associated with SBS often involve employee distrust of management. That distrust can grow from a perceived no-action response by management to a broad range of employee complaints.

In an atmosphere of perceived disregard and distrust, emotions cloud the situation. By the time a serious investigation is undertaken, it is hard to sort the legitimate complaints from those made in the hysteria of the moment. Recovering lost ground can be a long, painful, costly process.

Owner/tenant relationships can suffer the same consequence and the economic impact is apt to be more immediate and decisive. The tenant moves out. Word that the building is sick and the management is unresponsive spreads rapidly.

In too many instances, a "brushed off" complaint can lead to a very large public relations or legal problem even though the actual health problem may have been minimal. The air quality condition upon which the complaint is based may be minor, but the perceived indifference with which the complaint was received can explode into an issue beyond all recognition. Establishing procedures to respond to complaints is critical. A complaint response mechanism, because of its importance, frequently precedes any formalized IAQ program.

COMPLAINT RESPONSE PROCEDURES: Structure and Strategy

Step 1. Establish a complaint response procedure.

A critical aspect of establishing a complaint response procedure is briefing those, who will receive the complaints, on attitudes and procedures. Every potential complaint recipient should know that the cardinal rule is receive complaints with courtesy and careful attention. In other words, really listen. The person who complains may just want to be heard and to feel that someone cares. By listening and paying attention, a level of assurance is provided. Careful listening can also provide important clues that may lead to the solution of an underlying problem.

Complainants view all discussions of indoor air quality and possible treatments through a veil of personal experience. That veil cannot be pierced with talk of HVAC systems or the like until they have thoroughly exorcised their concerns, especially if the first complaint did not receive a personal response. They first must "unload" a litany of symptoms and any perceived neglect.

Formal notation of the complaint, and perhaps a formal interview later, not only gathers valuable data but sets the stage for more meaningful dialogue and interaction.

Step 2. Develop a complaint response tracking procedure.

A system for handling complaints of any kind should include a form for tracking and logging complaints. A complaint form and a log are essential management tools. In most instances, it is preferable not to have the complainant fill out the form, as the question format may actually suggest answers. It is usually better to assemble the information through a brief interview. (More extensive interview procedures are discussed in conjunction with investigations in Chapter 5.)

If several occupants have reported the same condition, or if the source of the problem is relatively obvious, it is not necessary to go through a description of the problem on each form.

The IAQ complaint form. The form need not be complex, but there are certain elements that should be included beyond the obvious date, name and route of contact used by the person who reported the problem. A form, similar to the sample complaint form shown in Figure 3-1, should be available to record of the complaint and initial actions taken. The demographic information and the types of questions should include:

- Complainant, work area, date received.
- Nature of the problem. The gross symptoms, pattern and location and times of occurrence.
- Any noticeable odors or abnormal conditions with descriptions.
- The name of the person taking the complaint and the person accepting the complaint for follow-up, if different, should appear on the form.
- Space should be included for a brief description of possible follow-up/remedial action to be taken. When follow up detailed investigations or extensive remedial actions are needed, these reports should be attached to a copy of the complaint form for filing purposes.

The Complaint Form

Building ___________ Floor ___________ Office/Room _____

Complainant ________________________________ Date ___/___/___

Nature of work ____

In this work location since ___/___/___ Problems since ___/___/___

Type of discomfort

Describe: ________________
____
____
____

Frequency of discomfort

________ times day, wk., mo.

Typical time ____________

Special conditions: ________
____

Environmental Conditions (circle, describe as appropriate)

too hot too cold too dry too humid

noticeable odor __________________________ time ________

noise ____

lighting ____

ambience/furniture ____________________________________

other ____

Outside Conditions (circle, describe as appropriate)

sunny partly cloudy cloudy rain storm snow

wind (strong/light) ________ outdoor temperature ___ °F

Special or Abnormal Conditions, Comments

____

____

Complaint recorded by ________________________________

Follow-up assigned to ________________________________

Disposition (Date & initial all entries)

____

Is further follow-up [ ] or monitoring [ ] warranted?

____

FIGURE 3-1

File # Assigned	Complaint Filed				Complaints/Conditions				Follow-Up			Comments
	Date	Bldg./Floor Office, Room	By	Com-plainant	Temp	RH	Vent	Other	Assigned To	Results	Initials	
001												
002												
003												
004												
005												

FIGURE 3-2. Complaint Log

Complaint forms are apt to be filled out by many different people in larger organizations. Any one with the authority to fill out the complaint, should be acquainted with all routing and disposition options.

It will defeat the purpose if complaints are allowed to sit on the corner of the desk for days. Getting the complaint to a central location promptly is a vital part of the process.

The complaint log. Once the complaint is received, it needs to be noted in a complaint log similar to the one shown in Figure 3-2. The log should provide space to note:

- the date received;
- who filed the complaint;
- who made the complaint and his/her work location;
- observed thermal environment and special conditions;
- the person designated for follow-up;
- disposition of the problem;
- if warranted, date and initials of any follow-up check or monitoring; and
- room for comments.

Step 3. Establish follow-up procedures.

Once the complaint is received and logged, it must be evaluated and some form of follow-up action taken. In many cases, the solution may be simple and quick. In others, a complaint may signal a problem that will require a great deal of detective work over an extended period of time.

If the investigation is protracted, it is important during this period that management make clear that complaints are taken seriously and that follow-up is real. This means that those who have complained and their co-workers need to be made aware that something is being done.

There is nothing wrong with overt action, even when it creates a slight disruption the work area. In the case of an odor problem, it pays to do some "sniffing" while people are at work; so they know investigators are on the trail of the culprit. Similarly, concern should be demonstrated. Ask questions and conduct the initial cursory examination of the ventilation system in the work place while it is occupied. This preliminary overt action is, of course, not just for show. Very often this relatively cursory investigation can provide early clues to the source of the problem.

Should a problem have medical consequences or require a detailed investigation, the diagnostic procedures will not be as simple, but the principle remains the same. Concerned workers need to know that the problem is getting careful attention and that measures are being taken to protect the occupants. If an investigation is underway, it is not a secret! It is best, therefore, to "tie a ribbon on it" and be open about what is being done. This avoids a flood of rumors and may be very useful in obtaining the cooperation of those who can provide valuable clues.

Step 4. Develop communication strategies.

Beyond the subtle and non-verbal communication potential available during preliminary investigations, a communications strategy must be an integral part of an emergency plan for any facility. It occupies a role no less important than meeting the challenge of a major power outage or the plan for evacuation. When conditions are less than ideal, effective communications can make the real difference between an inconvenience and a problem of enormous proportions.

Air quality problems may not require the speed of response of a boiler explosion, but communication with those involved is still a matter of high priority. Someone should have responsibility for the communications function within the emergency plan and should be fully aware of that assignment. This is not a task that can be left to "whoever is available." If the organization has the services of a professional public relations person, this assignment should be a part of the job description.

The collection and dissemination of accurate and timely information to the appropriate people is the cornerstone of a well-managed IAQ program. And the first place a troubled program "jumps the rails."

The #1 Audience. If there's a problem, those who occupy the facility must be the first audience. They need to be provided the facts, told what is going on and what is being done to rectify conditions. An audience that is well, and accurately, informed, greatly enhances the likelihood of keeping an indoor air problem confined to the facility. Failure brings about a much wider interest in the problem, including worker's families, the press and other outside groups.

Members of the #1 audience believe they have a real problem and, at least in their eyes, they deserve careful attention. They definitely should not feel their concerns are being ignored. Most facilities have at least one occupant who is never quite satisfied; so this is not always easy.

Communiques should convey that someone is following up on the complaint(s), that management knows and cares about their concerns and will keep them informed.

Getting The Word Out. When it has been determined that there is truly an indoor air problem, all the occupants of the facility should be made aware of the fact that management knows about it and is taking steps to solve it. How this is done will depend in great measure on the size of the organization, the occupants, and the seriousness of the problem at hand. In a school situation, for example, a potentially serious problem requires some communication with parents. In an office situation, such a broad effort may not be necessary. A simple posting or a "flyer" addressed to the occupants or tenant groups may suffice.

Communication to associated groups should be straight forward, stating facts as they are known, and setting forth the steps that have been taken and are contemplated. This should include the investigatory procedures and remedial actions to the extent they are known. It is important not to wait until all actions are "cast in bronze" before bringing occupants up-to-speed.

Rumor Control. Any organization with more than two people is fertile territory for rumors. And they will circulate at very nearly the speed of light. The most outlandish ideas will move as "actual facts" through any office, business, school or other facility where people get together. Rumors flourish in a "fact vacuum" making effective internal communications all the more vital. Rumors have a way of making the rounds even when the facts are easily available.

In the face of a serious indoor air quality problem, it may be desirable to designate an individual who can serve as a "rumor control center." Publicizing a telephone number where rumors can be checked and answers received will help curb rumors. It is absolutely necessary that the person delegated to handle rumors be kept fully informed of what is happening. If, through lack of current data, the wrong information is given out, the effort will be discredited and will backfire.

Progress Reports. Planned progress reports can head off problems. Let the occupants of the facility know each step of the way what is being done to "cure" the problem. If the solution will require a period of time, say so. Continued progress reports will help to control rumors and provide reassurance that management has not forgotten the affected individuals. This may be especially important in the case of public facilities.

Step 5. Develop Plans for Dealing with the Media.

In an area as volatile and emotionally laden as health problems caused by "unseen and uncontrolled villains" in our homes, offices and schools, the press plays a major role. Publicity has made more people aware that indoor air problems are a reality. Unfortunately, by repetition, over-coverage and failure to do sufficient homework, the press has been known to create a state of near panic where none was justified.

In some cases management has, in effect, allowed the press to diagnose a problem and prescribe a solution even before anyone could be certain that a real problem existed. This may not be the fault of the media, but a failure of management to communicate effectively with their own people or outside audiences.

Failure to acknowledge a problem when it exists and failure to explain what is being done to alleviate the problem leaves the way open for speculation and guessing from within and outside of the organization. Once false versions of what is occurring get started, they are virtually impossible to stop.

Projects do not always go smoothly or precisely follow a plan. Careful attention to communications throughout every phase provides the needed base of credibility if things start to go wrong. The key to successful passage through the communications mine field are openness, candor and adherence to the foremost rule of press relations: NEVER SPECULATE.

If problems arise, do not cut off the flow of information. Acknowledge problems while they are small and establish the fact that steps are being taken to deal with them. That way, hopefully, there will be no sudden revelation of a CRISIS.

When a serious (or sometimes not so serious) IAQ problem occurs in a public facility, it is fairly certain that the press will become involved. For those who have not dealt with reporters hot on the scent of a story, this can be an unsettling or frightening experience. But it need not be. Granted, it can be an intensely personal experience. Especially, the first time a person is misquoted in the press. So let's talk about the media and you.

The Media and YOU

There are a few basic rules to follow when dealing with the media that can take away much of the trauma and provide a much better chance that the newspaper or TV coverage of your problem will be accurate and fair.

Rule 1: Be accessible. If you have a problem, don't try to hide it. It won't work! Select one person to be the spokesman for the organization and be sure that they have all the information they need. Your spokesman should make sure that press queries are answered promptly. Press people work on rigid schedules. If your message is going to help, it must be timely.

Rule 2: Be open and honest. Tell them what you know and what you don't know. If an investigation is underway but no firm results are in, say so. Press people want instant answers to complex questions...they often know they can't get them, but they will ask anyway. Avoid giving preliminary information, because it may be wrong.

Rule 3: NEVER SPECULATE! (This is worth several rules.)
Speculation is telling something that you don't know. If the situation is less serious, speculation will amplify what could have been a minor problem. If the reverse is true, you have been guilty, or will be accused, of "covering up." Speculation is a sure way to lose when dealing with the media.

Answering a hypothetical question is a form of speculation. Don't succumb! Should a reporter's question begin with "IF," be on guard. Politely but firmly advise the reporter that you don't respond to hypothetical questions; then, turn the response to a point you want to make.

Rule 4: Admit you don't know. If you don't know the answer to a question asked by a reporter, say so. If possible, find the answer and get back to that reporter with the information. If you have given an answer, and you find out you were wrong, contact the reporter with the correct information. You may be too late to get the right answer in the paper or on the air, but you will be helping your credibility with the reporter and it will pay off over time.

Rule 5: You're on the record. Nothing you say to a reporter is ever "off the record." If you don't want to see it published, don't say it.

Rule 6: "No comment" is a comment. If you say, "I can't comment on that because we don't have the facts yet," that is an answer. A flat "No comment" is almost a challenge and may leave the impression that you know, but are hiding the facts.

Rule 7: It's YOUR answer. When being questioned in an interview situation, keep in mind that it is the reporter's question, but it is your answer. If you have a positive message that you want to get out, be sure to include it even if a leading question is not asked. It is perfectly all right to answer a question by saying, "I'll get to that in just a moment, but first I want to tell you about......."

Rule 8: Avoid pictorial amplification. Being interviewed in front of a catastrophe, or with investigators decked out in masks or "moon suits," magnifies the story and frequently sends the wrong signal.

Rule 9: Take some comfort. When all the news is bad and you seem to be all over the newspapers or the evening news, remember, most people will never hear about your problem. Sometimes it is a little hard on the ego, but it's reassuring to realize that everyone is self-centered enough not to pay too much attention to things that don't directly affect them. We are extremely aware of anything published about us, but everyone else is equally concerned with their own lives and may miss out completely on what grabs our attention. If you are taking what feels like a beating in the press, it generally isn't as bad as it seems.

MANAGEMENT BY COMPLAINT

The question that plagues management is how to determine whether a building is healthy or sick. Except for tobacco smoke, the threat cannot be seen and many pollutants are odorless. So what is the litmus test? Most of the literature starts with what Ylvisaker calls "the first and most obvious sign: tenant complaints."

While many owners and managers may get their introduction to indoor air problems through a complaint, relying on complaints leaves a lot to be desired. Using complaints as a barometer of indoor air quality poses two complications: (1) complaints do not necessarily mean the building is sick; and (2) the lack of complaints is no assurance the building is healthy.

In almost any facility at a given time, some occupants will voice complaints or evidence some of the symptoms typically associated with sick building syndrome. This natural occurrence can be blown into a full scale IAQ alert by an article someone found on office pollutants.

On the other hand, serious health consequences can emerge without evidence of symptoms or complaints long after the detrimental exposure has occurred. Problems associated with radon are a case in point.

If complaints serve as the trigger mechanism, an organization is automatically forced into a reactive mode.

Without a proactive approach to defuse unfounded complaints, the facility people are forever dashing about putting out "brush fires." This haphazard procedure places unexpected demands on manpower, usually when the organization can least afford to expend the time. Waiting "until the horse is stolen to lock the barn door" also leaves management in a legally vulnerable position. Failure to take appropriate measures, particularly if simple maintenance procedures would have avoided serious health consequences, is generally indefensible.

Once complaints are voiced, the longer it takes to identify the cause and to seek resolution, the more likely the performance, productivity and morale of everyone -- even management -- will be unfavorably affected.

After the office is a-buzz because of a "contaminant crisis" is a poor time to start learning what the legal requirements are. Finding time to learn, for example, what community notification requirements exist under the federal Hazard Communication Standard only adds to the chaos. In the long run, the "wait-and-react" procedures are more time consuming, more emotionally-draining and more costly.

Rask and Lane describe the typical management scenario:

> Building owners and/or managers may cut maintenance budgets without noticing any adverse effects for a period of time; even years. When problems do occur, such as SBS complaints, the managers may think that the people are complaining just for the sake of complaining. When the complaints continue, the building manager or responsible party will often hire a consulting firm to come in and "check" the air. We suspect that inadequate maintenance is rarely thought to be part of the problem.

The high cost of reacting to complaints in such a fashion gives new meaning to the old adage: An ounce of prevention is worth a pound of cure.

The advantages of a proactive program include providing occupants a comfortable productive environment; enhancing management/employee relations or management/tenant relations; improving building operations and maintenance; and providing an increasingly important selling point for marketing property.

Managing indoor air quality is not unlike other management functions. It entails setting a policy, establishing someone to assume responsibilities, and developing/implementing a plan. The greatest time will always be consumed in initiating the program. A well-orchestrated plan, once in place, can be essentially self-sustaining.

DEVELOPING AN IAQ PROGRAM

To avoid falling in the response-action trap, the first step is to develop a basic policy and make necessary protocol decisions. Unfortunately, the impetus for developing an IAQ program is frequently an IAQ "experience" that no one wants to go through again.

POLICY

The key function of a policy statement is to demonstrate a strong commitment by the governing board and/or top management. The policy usually incorporates a statement of concern and commitment as well as broad goals and enabling language. More specifically, it will include:

- a statement of concern stressing the importance of indoor air quality and the need to maintain a safe, healthy productive environment for the occupants;
- a statement of commitment by the board/management;
- the broad goals -- the conditions to be achieved and maintained;
- preliminary implementation considerations, such as;
  - —authorizing the position of IAQ manager, or designating such responsibilities to an existing position;
  - —delegating authority to that position within specified parameters, and
  - —requesting that an IAQ management plan be submitted for board and/or management approval; and
- reporting requirements, which incorporate evaluation data and further recommendations.

If the policy action has been prompted by a specific concern, the statement may make reference to this matter to reassure building occupants that top management is aware of, and committed to, resolution of the problem. For example, the direction to implement a plan may note respiratory problems being experienced by occupants in a particular area of the building and ask for a plan to respond to this need.

Other issues requiring top level support, such as new construction and purchasing procedures, might be addressed. If preliminary discussions reveal that a substantive shift in procedures or budget allocations will be required, a statement bolstering this change may be warranted. An examination of local IAQ conditions, for instance, may change operations and maintenance manpower needs in numbers, capabilities and training. Effecting this change requires resource allocations out of somebody's pocketbook.

Those giving up manpower and/or resources are inclined to create obstacles unless it is clear that top management is fully committed to the change.

The field of indoor air quality is young and new information is emerging rapidly. Some data will raise new concerns; other data, such as the 1989 asbestos reports from Harvard and the University of Vermont, will diminish old concerns. In this rapidly changing environment, policy statements must leave room for flexibility in their implementation. This can best be achieved if procedural specifics are not included in the policy, but left to the implementation plan or the judgment of the IAQ manager.

THE IAQ MANAGER

The term, IAQ manager, sounds formal and formidable. It is used only to underscore the truism: If everyone is responsible, no one is. Someone must be in charge whether it's a new position, or responsibility is delegated to existing personnel.

An effective IAQ manager needs to have technical expertise and leadership/communications skills. The available choices run the gamut from a truly fine engineer with limited communications skills, who understands an air handling system and can detect HVAC problems at a glance, to the eloquent leader, who wouldn't know a bioaerosol if it bit him. The ultimate answer obviously will vary with the organization's needs and other engineering or communications capabilities on staff.

Although an IAQ manager's job description needs to be situation specific, it typically will include responsibility for: (a) setting up and implementing an IAQ program; (b) establishing and maintaining IAQ records; (c) assessing IAQ needs, overseeing surveys and inspections; (d) making recommendations in conjunction with other facility personnel and/or a diagnostic team; (e) implementing approved recommendations; (f) writing specifications, procuring services and equipment; (g) overseeing installation and operation of the equipment; (h) planning and implementing internal and/or external communications strategies; (i) monitoring and evaluating the program's effectiveness; and (j) routinely reporting progress to board/management.

The IAQ manager's job description, qualifications and organizational placement will tend to flow from the tasks central to the job. Typically, those tasks will focus on resolving current IAQ problems or developing/implementing a management plan.

MANAGEMENT PLAN

Because of the diversity and complexity of managing an IAQ program, a plan is needed to outline the tasks to be performed. A notebook accompanying the plan, which contains protocols and more detailed procedures, can help treat specific needs in greater depth.

Depending on the size and nature of the firm, institution, or building, the plan is apt to vary in its formality, organization and presentation. An IAQ plan should be a living plan that moves with the times. If it sits on the shelf gathering dust, it was not responsive to needs at the time, or it was not flexible enough to meet changing conditions.

Most plans will include the following components, but not necessarily in the following order.

1. Person responsible for IAQ management. How, when, and where he or she can be contacted.
2. Purpose of the plan and how it is to be used.
3. Record keeping needs and procedures, such as
   - records required by law (which may have an overlap with "Right to Know" records),
   - complaint log and file of complaints/resolutions,
   - employees' record of exposures
   - findings from employee interviews; and
   - investigation reports.
4. Compliance requirements, with reference to
   - pertinent federal laws and regulations,
   - applicable state statutes and local ordinances, and
   - standards and guidelines; e.g., ASHRAE 62-1989.
5. Education and training procedures, which will cover
   - staff briefings,
   - the need for qualified personnel,

   - procedures to document worker qualifications through training and experience, and
   - any requirements for training attendance, certification, etc. of the organization's personnel.

6. Control and treatment procedures, including
   - general guidelines with reference to specific pollutants as discussed in the resource notebook,
   - safety procedures for O&M staff,
   - purchasing procedures,
   - time of use concerns,
   - special protocols; e.g., paints, pesticides,
   - temperature, humidity and ventilation parameters,
   - new construction/renovation design, commissioning requirements, and
   - storage of contaminants, disposal procedures.

NOTE: Some control issues may be listed separately in a plan, such as purchasing and new construction, because of their immediacy (new construction) or their importance to specific segments of the organization. If, for example, the procurement division is historically resistant to change, it may be necessary to pull out - and spell out - exactly what needs to be done.

7. Procedures to authorize and select outside consultants to
   - assist in establishing an IAQ program,
   - train personnel, and/or
   - conduct in-depth investigations.

8. Emergency preparedness procedures -- reference to a separate contingency plan.

9. Notification procedures including
   - any required notification to local, state and federal agencies,
   - notification to employees, including
     - postings, such as designated smoking or nonsmoking areas,
     - handouts that notify occupants of planned painting, spraying
     - information regarding exposures that have occurred, and

- notification to the community as required under the Hazard Communication Standard.

10. Complaint response procedures
    - filing complaints,
    - tracking and logging, and
    - follow-up.

11. Investigation protocol, including the
    - evaluation of healthy buildings,
    - interview forms, and
    - longer forms for data gathering to support in-depth diagnostics.

12. Communication strategies to keep employees and the community appropriately informed as to the
    - response to complaints,
    - investigations in progress,
    - remedial actions being taken, and
    - procedures for working with the media.

Concerns related to staffing, budget allocations and organizational procedures, developed as an adjunct to the plan, may be submitted as part of the plan or as a companion document.

The plan should be designed to satisfy the board's and/or management's concerns and to make the IAQ manager's job easier. The IAQ manager should be sure to put into the plan sufficient specifics to lend authority to his or her actions, but leave enough flexibility to meet unanticipated needs.

IAQ RESOURCE NOTEBOOK

The most functional approach is to accompany a plan with a resource notebook. A looseleaf notebook will permit easy revisions by section and the inclusion of new laws & regulations, articles, equipment data, etc.. The looseleaf approach facilitates revisions and helps keep the information current. The notebook can contain protocols and procedures in greater detail, such as the painting protocol discussed later in this chapter.

Information on each major contaminant or group of contaminants of interest can be assembled under dividers. (The material in Appendix A is organized so the segment on each contaminant might serve as a basic data sheet for these notebook sections.) Sections might also be devoted to purchasing guidelines, such as those available from American Conference of Governmental Industrial Hygienists (ACGIH), and new construction design requirements; e.g., ASHRAE 62-1989.

MANAGEMENT CONCERNS

As the management begins to cope with indoor air concerns, several issues usually emerge for consideration. These issues include economic matters, particularly in relation to energy costs; legal issues; and such specific policy decisions as smoking bans and painting protocols.

ECONOMIC FACTORS

All management decisions include some assessment of the cost/benefit ratio. What will a recommendation cost? What will it gain? Just how cost-effective is it?

Central to the benefit side of the equation is the owner's concerns regarding tenant turnover and unoccupied space; or management's worries about losses due to lower productivity and absenteeism. The financial losses and administrative nightmares to an owner of a sick building can be calculated. When the press reports a cost of $6 million to rectify the design/construction faults that lead to an uninhabitable facility, the costs to prevent such drastic action seem small in comparison. Costing out absenteeism is a little more difficult, as it may be clouded by other reasons for employee absence.

Lost productivity directly attributable to building associated health problems is hardest to measure. In 1989, respondents to a Building Owners Management Association (BOMA) stated that productivity would increase up to 21 percent if the air quality was improved. The BOMA survey reveals an estimate of the owner's perceived IAQ-associated productivity losses. At just 2 percent, the "IAQ" cost would run to $4 a square foot in an office building if total occupant cost runs about $200/sf/year. That cost doubles almost any preventive or remedial action needed to control most contaminants.

The Energy Cost Trade-Off

A cost generally associated with improved indoor air quality is the utility costs for increased ventilation. Should energy costs climb in the coming years as projected, these costs will mount dramatically and ventilation as a control measure will come under greater scrutiny.

The dilution solution has been given impetus through the revised ASHRAE standard 62-1989, "Ventilation for Acceptable Indoor Air Quality." While 62-1989 does offer an indoor air quality procedure for judging ventilation needs, its "Table 2" (cfm/occupant) offers a guideline, which most engineers are apt to use to calculate outside air needs. Unfortunately, the ultimate impact of 62-1989 will probably cause cautious managers to increase outside air three to four times in healthy buildings and/or under conditions where quantity cannot equate to quality. ASHRAE's 62-1989 is discussed in greater detail in Chapter 10.

Operating losses through absenteeism and lower productivity can far outweigh any energy savings consideration. On the other hand, using ventilation to satisfy health needs does not come cheap. Ventilation proponents suggest meeting the new ASHRAE 62-1989 cfm/occupant guidelines will increase energy costs by 5 percent, citing Eto and Meyer's study. The building conditions in their study, however, are not generally typical of most buildings. Increased cost projections for increasing ventilation from 5 cfm to 15 cfm have run as high as 32 percent; and to 20 cfm, over 50 percent.

Careful management need not sacrifice energy savings for air quality. Energy efficiency and indoor air quality do not have an adversarial relationship, but rather are part and parcel of the comfortable, productive environment facility managers seek to deliver. Owners and managers need to understand the limitations of ventilation as a solution, as discussed in Chapter 8, as well as the ways engineers can address indoor air quality, which will satisfy ASHRAE 62-1989 without automatically turning up the fan.

As energy prices climb the need to recognize the options available and their cost-effectiveness will also grow. In the face of such a scenario, it is well to remember that in its 1989 report to Congress, the EPA concluded that the best control for any contaminant is at its source -- not through dilution by ventilation.

LEGAL MATTERS

Potential legal costs become a part of any economic consideration. Legal fees and judgments in today's litigious society can weigh very heavily on budgets and management time. Attorney/engineer Gardner has warned, "Indoor Air Quality will be the toxic tort of the 90s." One of the best reasons for developing a proactive IAQ management program is the ability to demonstrate a "reasonable man" position in order to assume the protective mantle that the doctrine affords.

SPECIFIC POLICY ISSUES

Some "policy" issues will be part of the policy statement, others may be part of an approved plan, and still others will be left to administrative judgment. How these issues are handled will depend on organizational size, management style and how much administrative "weight" is needed behind them. Concerns related to smoking and painting are excellent examples.

The need for enforceable management decisions is central to environmental tobacco smoke (ETS) control strategy. Most ETS remedies require some control of occupant behavior through bans and designated smoking/nonsmoking areas. Such decisions, once made, must have the full force of the board/administration and top management to be effective. In radon affected areas, the medical, economic and legal implications take on added import as the radon/ETS synergism increases the likelihood of lung cancer significantly.

Some special protocols developed and circulated in advance, such as painting procedures, can very easily ward off a contaminant problem. The selection of the paint, the time of application, the advance notification to chemically sensitive occupants, and procedures to purge the facility prior to re-occupancy are all matters that can be resolved in advance. A large school system, Anne Arundel County Public Schools in Maryland, has developed and implemented a comprehensive paint protocol. This protocol is presented in Appendix B as an example of how management can anticipate certain concerns and associated indoor air problems can be avoided.

SECURING CONSULTANT SERVICES

Much of IAQ management involves solid preventive maintenance and common sense. Technical needs can often be met with internal staff; particularly, if the staff has received some IAQ training. There usually comes a time, however, when management needs outside assistance, especially in the planning stages or in diagnosing complex IAQ problems. When the time comes for outside help: be careful of the "snake oil salesman."

As in any relatively new field, there are those who would like to profit from the client's ignorance. They can often be spotted by:

- their propensity to use IAQ jargon and meaningless statistics;
- a general reference to all the work they have done, but a failure to be specific;
- scare tactics; the "You-are-living-in-a-time-bomb!" approach, or emotional blackmail; e.g., letters of warning mailed directly to building occupants, or in the case of schools, to the parents;
- a product, a new magical "black box," that will take care of all your indoor air problems; and/or
- a guarantee to discover the cause of your indoor air problem(s) before ever stepping in the facility.

Chapter 5, Investigating Indoor Air Problems, discusses technical concerns related to retaining investigators and suggestions for using them effectively.

To avoid those who would jeopardize the organization's IAQ efforts, consultants, registered professionals or certified technicians should be hired in the same manner as any other professional: ask for and check credentials and references.

EDUCATION AND TRAINING

In addition to the communications strategies associated with complaint response discussed earlier in this chapter, the management has internal education and training responsibilities, including staff briefings and employee training. Some of these responsibilities may be fulfilled by the organization's own personnel; others may require an outside consultant.

Increasingly, many activities associated with an IAQ program are covered by laws and regulations. Ignorance of these provisions does not exonerate the individual or the organization from compliance. Legal precedents have been established for prosecution on the basis that the defendant "should have known." Briefings and training, therefore, have become a key IAQ management function. Some training is specified by law, while additional training can afford individuals and the organization protection against costly mistakes or criminal liability. IAQ is a rapidly emerging field of concern at the federal and state levels; so checking with state agencies periodically is a good practice.

Staff briefings. All staff need to know the purpose behind the IAQ program and the general procedures for implementation. They should be given guidance in reporting problems in a context that does not "manufacture" IAQ concerns or push them out of perspective.

Laws, regulations and standards that have implications for all employees need to be a part of the staff briefings and made available to them. Such documents should be part of the resource notebook.

Middle management and supervisors need to know what is expected of their workers and exactly what steps should be taken in case of an emergency. They should be advised of operational procedures, requirements under the law and their respective roles. As noted earlier, particular care needs to be taken so they understand information dissemination procedures to staff, public and the press.

If unions are involved, the union leadership should also be briefed, as appropriate, on the IAQ program and procedures.

Employee training. Specific "hands-on" training should be provided to prepare workers to perform their tasks safely and effectively. This training should encompass: (1) the safe use and storage of hazardous materials; (2) how to perform their duties to avoid indoor air problems; (3) how to deal with any IAQ problems that may arise; and (4) procedures to assure air quality.

As stressed in the maintenance chapter, preventive maintenance (PM) is the centerpiece of an IAQ program. A PM training program with appropriate support materials should be developed and conducted to assure effective program implementation.

COMMUNICATIONS IN IAQ MANAGEMENT

Extensive communications relative to indoor air quality is not a major issue until "THE PROBLEM" emerges. Communications then tend to take on a level of importance that may well overshadow the real IAQ problem.

But, to limit concern for communication to times of crisis is a serious mistake that can have costly consequences. A continuous flow of information plays a vital role in any ongoing program to insure the quality of the indoor environment.

Effective IAQ management means that key people must understand and fully accept their role in all phases of the effort. Effective communications is the tool that helps to achieve that understanding and becomes as much a part of successful program as the wrench and screwdriver.

IAQ information needs to be presented to management in a straight forward manner. Burying people with detailed discussions of bioaerosols, VOCs or information about specific pieces of equipment will only confuse the issue. A general discussion should focus on what can be done, the problems that can be avoided, dollars that can be saved, and the comfort as well as health protection that can be achieved.

As an important side benefit, it has been demonstrated over and over that effective internal communications is one of the most effective external communications tools. The staff talks to friends and the word spreads.

Effective communications at all levels allows an IAQ problem to be handled in an atmosphere that permits an orderly approach, without panic. Good communications requires a small investment of time and resources, but the management dividends that accrue can be enormous.

In summary, effective IAQ management must be adapted to the organization's needs and capabilities. Whether an organization becomes IAQ sensitive because it has a problem or wants to avoid one, the inescapable components of an effective IAQ program are the statement of commitment, the designation of responsibility, identification of the technical and financial resources to do the job, a carefully designed communications strategy and a workable plan, which includes an effective complaint response procedure.

Chapter 4

Classifying Indoor Air Problems

Owners and managers seldom have the luxury of knowing the full range of pollutants in their buildings. If they have an indoor air problem, they hear the complaints, symptoms, or the medical diagnosis. But they aren't handed a list of contaminants.

Unfortunately, most of the IAQ literature is based on specific contaminants. That's not particularly useful to the practitioner who is confronted with an indoor air problem that needs attention now with no idea what pollutant is causing the problem.

The answer would seem to be: figure out what contaminant is causing the problem. But it's not that easy. Identifying contaminants often proves to be difficult, costly, even impossible. The situation can be further confused by the multifactorial nature of indoor air problems. Even if a contaminant in the area has been identified another undetected contaminant may be the one at fault. Or, they may be acting synergistically to cause the problem.

Working from the manager's perspective, symptomatology and diagnosis offers a preliminary classification opportunity. It offers a way to get a handle on the situation from what is already known. This chapter, therefore, looks at the situation as it is first made known to management. We start with the complaints, the symptoms; only then, do we look at major contaminants and environmental conditions to explore their relationship to these health effects.

SICK BUILDING SYNDROME AND BUILDING RELATED ILLNESS

Sick Building Syndrome (SBS) is a world-wide problem. SBS initially was used to describe a building where a set of varied symptoms were experienced predominantly by people working in an air conditioned environment. Subsequent study by Finnegan and others has shown that SBS is not limited to air conditioned facilities and can, in fact, be observed in naturally ventilated buildings.

A syndrome, by definition, is a group of signs and symptoms that occur together and characterize a particular abnormality. Frequently they form an identifiable pattern. This makes diagnosis by exclusion possible. For example, organic lesions are not associated with SBS. If an occupant has persistent organic lesions, it can be assumed that the cause of the sickness is not SBS.

As defined in Chapter 1, a building is said to be "sick," when 20 percent or more of the occupants voluntarily complain of discomfort symptoms for periods exceeding two weeks, and affected occupants obtain rapid relief away from the building. The 20 percent figure, derived from earlier ASHRAE efforts to define comfort, may mislead managers who look to "20 percent" as a guideline for action. As Michael Crandall of NIOSH has observed, "If you have 3,000 employees and only 10 percent are ill, do you wait? That's 300 people who could be suffering needlessly from a pollutant in the work place."

A problem closely associated to SBS is building related illness (BRI). Medical diagnosis can identify specific health effects, such as Legionnaire's disease, that are a direct result of building conditions. Once diagnosed, a BRI can help identify the source and may reveal ways to remedy the situation. A BRI facility has almost always passed through the SBS stage and usually still has other contaminants at work.

The five symptom complexes associated with SBS are discussed below; followed by the major building related illnesses.

SBS SYMPTOMATOLOGY

The five SBS symptom complexes can occur singly or in combination with each other. Symptoms may be cyclical or episodic. They may be nonspecific and often resemble a common cold or other respiratory illnesses. They usually get worse as the day progresses, may worsen through the work week, and ease up or disappear once the occupant is away from the building for a time.

1. Eye Irritation

A burning, dry, gritty sensation is experienced in the eyes without any evidence of inflammation. Severity varies from day to day. Sensitivity is greater for occupants wearing contact lenses.

2. Nasal Manifestations

The most frequently cited nasal symptom is "stuffiness," which develops rapidly when an individual enters the building, persists while in the building, and goes away quickly upon departure. For some people, this "stuffiness" is a specific reaction to high temperatures. Other nasal symptoms, which are more variable and apt to be less persistent, are nasal irritation and rhinorrhoea. Symptoms are frequently suggestive of an allergic cause.

3. Throat and Lower Respiratory Tract Symptoms

A persistent dryness of the throat, which seldom shows any inflammation, is a principle symptom. The occupant may gain some relief by drinking large amounts of water.

A typical indication of lower respiratory tract difficulty is a shortness of breath, a sense of not being able to breath deeply, which is not related to any lung infection or bronchial asthma. It is generally relieved by stepping outside to take a few deep breaths.

4. Headaches, Fatigue, General Malaise

The headaches are usually frontal in position, occur in the afternoon and may occur daily. Headaches may range from moderate to severe migraine.

Headaches, fatigue, dizziness, difficulty in concentration and general malaise are the most frequently cited sick building symptoms.

5. Skin Problems

Dry skin is a frequent SBS complaint, particularly from female occupants. It is considered a building associated symptom when it improves during protracted absences from the facility. Warm dry air or excessive air movement may create a particular type of dermatitis on exposed skin surfaces. Skin rashes may result from exposure to some contaminants.

SBS may aggravate existing health problems and diseases, such as sinusitis and eczema, but these are considered outside the general SBS symptom complexes.

BUILDING RELATED ILLNESS DIAGNOSIS

Building associated sicknesses, other than those related to SBS, are generally allergic reactions or infections. The allergies include asthma, humidifier fever and hypersensitivity pneumonitis. Bacteria, fungus and virus cause the BRI infections. While BRIs have some characteristics in common, there are others, as noted below, that help distinguish one from another. Careful documentation of symptoms and complaints as well as their patterns, and predisposing conditions can be used as tools to focus further diagnostic activity.

Allergies

Asthma, rhinitis. Symptoms are those generally associated with asthma and increase in intensity with prolonged exposure; i.e., a work week, but improve away from the facility.

Allergic responses of the upper and lower respiratory tracts generally are secondary to the inhalation of allergens. These allergens are usually associated with poorly maintained buildings and may originate in humidifiers, particularly cold water spray humidifiers, contaminated by microorganisms.

Hypersensitivity Pneumonitis. This generic term for a common manifestation, is also referred to as extrinsic allergic alveolitis. It is basically an allergic lung disorder and it occurs in varying intensity. After a week to ten days, symptoms usually regress without further exposure. Symptoms get increasingly worse during exposure and show some relief when causal agent is absent, such as a week-end. Malaise and myalgia (muscular pain) are almost always present. Headaches frequently occur. In the more acute phase, shortness of breath, fever, chills and a dry cough appear within an incubation period of about six hours. Attack rate is very low. Individual susceptibility is suspected as a key factor.

Humidifier Fever

Symptoms of humidifier fever are "flu-like," lethargy, arthralgia (neuralgic pain in the joints), myalgia and fever. Sometimes headaches, polyuria and weight loss occur. More severe symptoms include shortage of breath and coughing. These systemic and respiratory symptoms occur with initial exposure; i.e., the first day of the work week. They progressively improve during exposure and in the absence of exposure, only to recur on re-exposure. In a work setting, symptoms appear on Monday, improve during the week and weekend and recur the following Monday. This pattern clearly distinguishes humidifier fever from hypersensitivity pneumonitis.

During the height of the reaction, medical examinations reveal the presence of late inspiratory crackles during auscultation ("listening" to the chest) and impaired gas transfer in the lungs. The chest radiograph is always normal and the lung function is normal between attacks.

The cause or causes of humidifier fever are not known. Some organisms have been isolated and incriminated during outbreaks, but as yet not indisputably identified. However, immunological investigations almost always reveal the presence of precipitating antibodies to antigens extracted from the humidifiers. In bronchial provocation tests, water from the humidifier usually reproduces the symptoms and physiological changes.

Attack rates vary considerably in reported outbreaks. Age appears to be a factor and appears to be associated with the duration of exposure and the development of antibodies. Highest incidence rate is in the winter.

In addition to the tolerance pattern, or the "Monday Morning Phenomenon" of humidifier fever, it is distinguished from hypersensitive pneumonitis in other ways. Humidifier fever shows no decrease in lung function or pulmonary fibrosis, there are no radiographic changes and it seems to be brought on by comparatively low levels of antigen whereas hypersensitivity pneumonitis is associated with massive antigen exposure.

Infections

Bacterial. Legionnaires' disease, caused by Legionella pneumophila, is the most widely recognized bacterial form of BRI infectious disease. Features typical of Legionnaires' diseases are headache, chest pain, vomiting, diarrhea, weight loss, fever, dry cough, recurrent chills, dyspnea, myalgia, abdominal pain and pneumonia. It has a 15 percent fatality rate.

Legionnaires' disease is not a recent phenomena and has been backdated to 1947. While occasionally appearing in epidemic proportions in certain environments, it is more apt to be sporadic in nature. Occurrence has been estimated as high as 116,000 cases per year.

The incubation period for Legionnaires' disease at onset is two to ten days and the attack rate is a low 6 percent. It generally favors males and individuals over 55. Smokers are roughly 2 to 5 times more likely to get the disease. It appears to have a summer-fall seasonality.

Pontiac fever is a relatively mild clinical form of Legionnaires' disease. Often related to specific worker activity it can occur with a high attack rate from a common source, such as cleaning steam turbine engines. Fever, malaise, headache, chills, myalgia, nausea and diarrhea are also features of Pontiac fever. Respiratory symptoms, such as sore throat and a slight cough, may be present, but pneumonia is not associated with the disease. This nonpneumonic form of Legionnaires' disease has a shorter incubation period, 5 to 66 hours, and an almost 100 percent attack rate. No age or sex distinction is evident. Outbreaks occur in summer. No fatalities have been reported.

Medical diagnosis of Legionnaires' disease and Pontiac fever can take several weeks, even months, as the necessary serologic diagnosis requires the appearance of antibodies.

Fungal and viral infections. The list is long, as over 100,000 fungi have been described and up to 2,000 are added each year. Infections, for example, have been caused by fungus, aspergillus, coming into the buildings from outside air and contamination in the duct work has been well documented. Of particular concern are old or immunocompromised patients in hospitals. Some fungi produce volatile organic compounds; so VOCs can be of biological origin.

With the greater attention indoor air quality is receiving, it is easy to overlook the viral infections. This area has been well treated in the medical literature and is not addressed here. It is interesting to note, however, that an air-conditioning system appeared to be the means of spreading an epidemic infection of measles in a school.

Table 4-1 summarizes health symptoms and associated SBS and BRI illnesses. The table is particularly useful in identifying which illness is the only one to exhibit a certain symptom, or only illness that does not exhibit that symptom. This helps with the sorting process. For example, eye irritation is only identified with SBS; not BRIs. Humidifier Fever is the only illness to exhibit lethargy but not general malaise. Malaise is a vague feeling of uneasiness or physical discomfort. Lethargy is characterized by abnormal drowsiness or torpor, apathy, sluggishness and great lack of energy. When a building associated illness is suspected, then, abnormal drowsiness without any feelings of physical uneasiness would suggest Humidifier Fever. If these symptoms are accompanied by the "Monday Morning Phenomena," the possibility of Humidifier Fever is even greater.

Table 4-2 offers some key factors that help distinguish Building Related Illnesses (BRI) from each other. Certain factors, such as seasonality, help identify the illness. Other information, such as predisposing factors, suggest situations where preventive action may be more critical.

TABLE 4-1. HEALTH SYMPTOMS AND ASSOCIATED SBS AND BRI ILLNESSES

ILLNESS	Health Symptoms																		
	Chest Pain	Chills	Concentration Diffl.	Cough	Dizziness	Eye Irritation	Fatigue	Fever	Headache	Lethargy	Malaise	Muscle Ache	Nausea	Pain in Joints	Pheumonia	Shortness of Breath	Skin Irritation	Weight Loss	Other
SBS			■		•	■	•	•	•		•					•	■		Dryness of throat nasal stuffiness rhinorrhea
Legionnaire's Disease	•	•		•				•	•		•	•			■	•		•	Abdominal pain, confusion, diarrhea, vomiting
Humidifier Fever	•	•		•				•	•	■		•		■		•		•	Polyuria
Hypersensitivity Pneumonitis		•		•			•	•	•		•					•			
Pontiac Fever	•	•		•	•			•	•		•	•	■						Diarrhea, sore throat

• Symptom regularly occurs with the illness.

■ Only building related illness where symptom typically appears.

TABLE 4-2. BUILDING-RELATED ILLNESSES *

BUILDING RELATED ILLNESS	ATTACK RATE	INCUBATION PERIOD	SEASONALITY	PREDISPOSING FACTORS	RELATED COMMENTS
Hypersensitivity Pneumonitis	3-16%	6 hours	—	No apparent disposing factors	Only small proportion of exposed develop active disease
Humidifier Fever	Variable 2.6 to 70%	4 hours	winter	SM: negative correlation between smoking and presence of antibodies Ex: duration of exposure, development of anti-bodies and age	"Monday morning phenomenon"
Legionnaires' Disease (LD)	6%	2-10 days	summer - fall	Age: Above 55 years Sex: Males SM: more common in smokers	Avg. 15% fatality rate Pulmonary involvement More common in those with underlying disease
Pontiac Fever	95-100%	5-66 hours	summer	Age: Lower than LD but reflects age distribution of exposed Sex: reflects sex distrib-ution of exposed (but high attack rates)	No fatalities Lack of pulmonary involvement

* Adapted from work by J.M. Benard, Montreal Canada

It is important to recognize that many of the SBS and BRI conditions described above may not be caused by the building. A sudden outbreak of flu-like symptoms may, in fact, be the flu and not humidifier fever.

PUTTING HEALTH INFORMATION TO WORK

Because epidemiologists, engineers, architects, physicians, industrial hygienists, researchers and regulators all have their own approach to addressing indoor air quality, each group has emphasized different aspects and the relative importance of various diseases. One of the great weaknesses in indoor air quality research and problem resolution available to practitioners is the absence of discussion and consensus among the disciplines involved.

Others have grouped building associated maladies in order to define sick buildings, or the rate or prevalence of certain symptom patterns. Early work by Robertson et al. grouped the symptoms as dry symptoms, allergic symptoms, asthma symptoms, and uncertain cause. Hodgson and Kreiss using an epidemiological approach, developed the following groupings; allergic respiratory disease, mucous membrane irritation, infections, dermatitis, reproductive complaints, miscellaneous, and tight-building syndrome.

In an effort to relate symptom clusters to the type of building ventilation, Hedge et al. established three general factors; general health, mucous membrane, and musculature. In their work, they found no "mucous membrane" symptom clusters in naturally ventilated buildings. The results of their research also revealed that symptoms, such as skin dryness, fever and respiratory problems, did not form significant clusters even within air-conditioned buildings.

Clearly no hard and fast grouping of symptoms can unequivocally establish a medical paradigm. The paradigm used in this chapter most closely follows the system complexes associated with SBS and the specific BRIs used by the Joint Research Centre - Institute for the Environment of the Commission of the European Communities.

It should be stressed that most of the research reported in the literature, upon which this examination is based, does not meet stringent scientific criteria. Information on the investigation of healthy buildings, for example, is woefully lacking. Few studies have used "control" buildings. When used, efforts to establish controls as representative of the population of control buildings has not been clearly documented. The work by Theodor D. Sterling and Associates, Ltd. and Healthy Buildings International, Inc. to set protocol for indoor air quality monitoring for the City of Columbus IAQ project is a much needed addition to the field of IAQ research.

In whatever fashion complaints and symptoms are grouped, attributing them to contaminants and building sources, wherever possible, is of critical importance. Table 4-6, at the end of this chapter, lists the symptom complexes and diseases with probable causes by contaminants, environmental conditions and primary sources. To make full use of the table, some knowledge of the probable causes, both contaminants and conditions, is important.

CONTAMINANTS AND THEIR SOURCES

Contaminants may originate outside the building to be transported or ventilated into the facility, or they may be generated in the building. While in the building, they have a limited number of options available to them; The indoor pollutant flow shown in Figure 4-1 provides an overview of the contaminant's "life" in a building.

Indoor air quality is determined by a range of conditions and the interactions of sources, sinks and air movement among rooms and between the building and outside. Sources as depicted in Figure 4-1 may be located in building materials, furnishings or the HVAC system. Other sources are consumer products and office equipment. The occupants themselves constitute a major source of pollution, especially smokers. Sinks may be located in the rooms or systems and may ultimately become sources themselves. Air movement in a building consists of (a) natural air movement among rooms sometimes fostered by occupant movement, (b) air movement driven by a forced air system (HVAC); and air movement between the building and outside through ventilation, infiltration and exfiltration.

As noted earlier, most discussions of indoor air quality are discussed from the perspective of specific pollutants. Usually an attempt to organize the pollutants in some fashion is made. As with symptoms, the organizational patterns seem to be as diverse as the disciplines doing the work. Early efforts by Woods sorted pollutants by physical characteristics, "mass" and "energy." A few have sought to address contaminants by their source; e.g., those entrained in a building, combustion products, activities of humans.

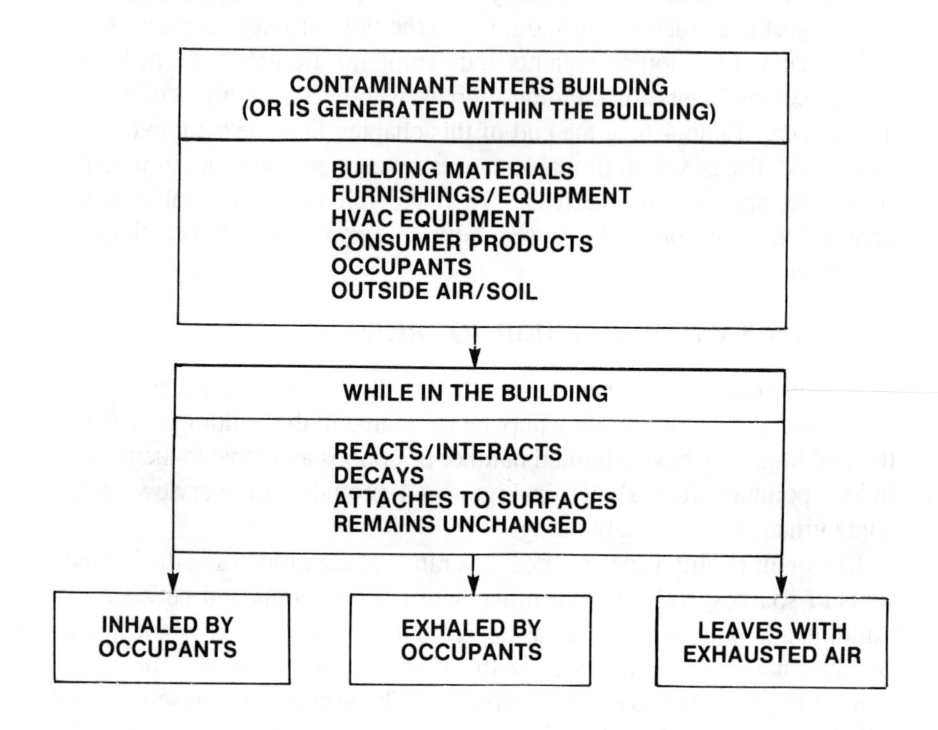

FIGURE 4-1. INDOOR POLLUTANT FLOW

The composition of chemicals known collectively as indoor air pollutants is a complex array of gases, vapors and particles. The determination of health effects related to these pollutants collectively, individually, or in certain combinations requires extensive information about the exposure of an individual to this mixture. A body of literature is developing that provides information in varying degrees on the major indoor pollutants that affect human health. These pollutants can be divided into particles (solids and liquid droplets) and vapors and gases.

Particles of specific interest include:

1. Respirable particulates as a group (10 microns or less in size);
2. Tobacco smoke - solid and liquid droplets - as well as many vapors and gases;
3. Asbestos fibers;
4. Allergens (pollen, fungi, mold spores, insect parts and feces); and
5. Pathogens (bacteria and viruses), almost always contained in, or on, other particulate matter.

Vapor and gases of particular interest include:

1. Carbon monoxide (CO);
2. Radon (decayed material becomes attached to solids);
3. Formaldehyde (HCHO);
4. Other volatile organic compounds (VOCs); and
5. Nitrogen oxides (NO and NO_2).

Increasingly, the procedure has been to treat the major indoor pollutants individually or in small groupings; e.g., bioaerosols, which includes allergens and pathogens.

Since the practitioner is most apt to be concerned about the relationship of particular health manifestations to particular contaminants or practical groupings, the latter approach is employed here. In this way, the facility manager, who is wondering if wet carpeting in the lounge could cause asthmatic reactions can turn to bioaerosols. Or, the plant engineer, who may be trying to discern a relationship between the HVAC leaks he has just detected and coughing and wheezing among occupants, will find the approach workable. Certainly those in a facility responsible for providing a safe comfortable productive environment worry about tobacco smoke as an entity and not about the 4700 chemical compounds that come from a lighted cigarette.

The reader should be aware that none of the classifications, including the one used here, is totally satisfactory, as there is considerable overlaps among individual substances. Tobacco smoke, as just indicated, is an obvious example. The joint effect of sidestream and mainstream tobacco smoke to the passive smoker is to expose him or her to both particulates and gases, including a long list of volatile organic compounds. The organization of contaminants on the following pages then, was predicated on the most useful treatment for the practitioner, but many of the pollutants as listed are not mutually exclusive.

In this chapter, the description, the probable sources, the symptoms and health effects are discussed for each major contaminant. Later chapters discuss treatment and control procedures as well as laws/regulations/standards. References for contaminants throughout the book are all brought together for each contaminant and placed in Appendix A. For the reader, who wishes to pursue a particular pollutant in greater depth, selected references and primary sources of information are also provided with the summary of each contaminant.

ASBESTOS

Asbestos is a term used to refer to a number of inorganic minerals that have specific properties in common. The serpentine mineral, chrysotile, is the most commonly used asbestos and represents about 95 percent of the asbestos used in buildings in the United States. The second largest asbestos group is amphiboles, which includes amosite and crocidolite. The fiber structure and the associated health risks are very different by type.

Once in place, asbestos does not degenerate spontaneously. Fiber bundles are not inclined to be disrupted without some mechanical external force.

Sources. Man has used asbestos for thousands of years. Until the 1970s, it was the material of choice for thermal and acoustical insulation as well as fire-proofing. It can be found in thousands of commercial products including reinforced cement and heat-resistant textiles.

In the building structure, asbestos containing material (ACM) is most frequently found in boiler insulation, pipe insulation, sprayed-on fire proofing, breaching insulation, floor and ceiling tiles. The EPA estimates that 20 percent, or 733,000, buildings in the country contain friable asbestos — and this figure excludes schools and residential buildings with less than ten units.

Symptoms and health effects. There are no immediate discernable symptoms of asbestos exposure. EPA and medical specialists have estimated that 1,000 to 7,000 already exposed people will die of asbestos-related diseases over the next 30 years.

All current information on the health effects resulting from asbestos inhalation come from studies of occupational settings with high exposure levels. Most data are based on the amphiboles group, not common in U.S. construction. Three forms of disease have been associated with the inhalation of asbestos fibers: (1) asbestos-fibrosis or scarring of the lungs; (2) mesothelioma - a malignancy of the linings of the lung and abdomen; and (3) lung cancer.

Asbestosis - deaths have only been observed among individuals that have been occupationally exposed to high levels of asbestos. There does not appear to be any evidence that asbestosis should be a concern as a result of environmental exposures today.

Mesothelioma and lung cancer - conservative assessments by various researchers, as indicated in the Harvard report, Symposium of Health Aspects of Exposure to Asbestos in Buildings, place the associated lifetime risk of death at 1 per 100,000 for 10 years of occupancy in buildings with 0.001 regulatory fibers/ml of mixed fiber types. Recent data suggest average level of asbestos in schools and other buildings with ACM is generally well below 0.001 fibers/ml. The Harvard researchers observed that the lifetime risk of one premature death per 100,000 associated with asbestos exposures was small compared with other environmental risks. The data in Table 4-3, taken from the Harvard report, offers some comparative risks by cause.

TABLE 4-3. PUBLISHED ESTIMATES OF RISK FROM VARIOUS CAUSES (Mainly U.S. Data)

CAUSE	LIFETIME RISK OF PREMATURE DEATH (PER 100,000)
Smoking (all causes)	21,900
Smoking (cancer only)	8,800
Motor Vehicle	1,600
Frequent Airline Passenger	730
Coal Mining Accidents	441
Indoor Radon	400
Motor Vehicle - Pedestrian	290
Environmental Tobacco Smoke/Living with a Smoker	200
Diagnostic X-rays	75
Cycling Deaths	75
Consuming Miami or New Orleans Drinking Water	7
Lightning	3
Hurricanes	3
Asbestos in School Buildings	1

Source: Harvard report, Summary of Symposium on Health Aspects of Exposure to Asbestos in Buildings. Based on Sources of Risk Estimates: Commins (1985), Weill and Hughes (1986), Wilson and Crouch (1982)

The primary health risk among occupants would appear to be to operations and maintenance people who, in the course of their duties, may disturb ACM. The greatest health risks are, of course, to those who remove the material. OSHA guidelines for asbestos removal should be followed to minimize this risk.

Mounting evidence that asbestos is not as great a problem as previously surmised prompted Congress in 1989 to request the EPA through the Health Effects Institute to reassess asbestos-related health risks.

BIOAEROSOLS

Bioaerosols, or airborne biological agents, include fungi (yeasts and molds), dander, spores, pollen, insect parts and feces, bacteria and viruses. Bioaerosols are also discussed in literature addressing microbiological, or microbial, contaminants.

Microorganisms associated with normal childhood diseases, such as mumps and measles, or the usual afflictions like flu and "colds" are not normally treated in IAQ literature.

Sources. Biological growth sources include wet insulation, carpet, ceiling tile, wall coverings, furniture and stagnant water in air conditioners, dehumidifiers, humidifiers, cooling towers, drip pans and cooling coils of air handling units. People, pets, plants, insects may carry biological agents into a facility or serve as potential sources.

Biological agents may enter the building through outside air intakes. Due to their small size, they may not be filtered out of the airstream. Frequently, they settle in the ventilation system itself. Fiberglass insulation, often used to replace asbestos, can provide a fertile territory for bioaerosols. The inevitable dust and darkness inside duct work, plus condensate moisture, all work together to turn the vast surface area of fiberglass into breeding grounds for mold.

Symptoms and health effects. Common symptoms associated with biological contaminants include sneezing, watery eyes, coughing, shortness of breath, dizziness, lethargy, fever and digestive problems.

Different causes prompt diverse symptoms and conditions. Concerns may range from odors and "stuffiness" to Legionnaire's disease. No evidence presently exists to support the contention that bioaerosols are responsible for medical problems related to skin or reproductive systems.

Building related illnesses (BRI) constitute a range of hypersensitivity and infectious diseases. Hypersensitivity diseases, such as asthma, humidifier fever and hypersensitivity pneumonitis, are caused by immunological sensitization to bioaerosols. Prolonged exposure to mold spore allergens, for example, wears down the immune system and increases the individual's sensitivity prompting reactions to lower concentrations. One case of hypersensitivity pneumonitis is sufficient to suggest the sensitization process may be occurring in others, and mitigation procedures should be taken. In any suspected building-related illness, a physician should be involved in the diagnosis and the etiology.

If a health complaint involves an infectious disease; e.g., hypersensitive pneumonitis or Legionnaire's disease, it can be assumed that a bioaerosol is involved.

COMBUSTION PRODUCTS

The major categories of products resulting from incomplete combustion can be listed as carbon monoxide (CO), nitrogen oxides (NO_x), particulate material, and polynuclear aromatic hydrocarbons (PAH).

Carbon monoxide (CO) is an odorless, tasteless and colorless gas.

Nitrogen oxides (NO_x) includes nitrogen compounds NO, NO_2, N_2O, OONO, ON(O)O, N_2O_4 and N_2O_5. All are irritant gases, which can impact on human health.

Particulates represents a broad class of chemical and physical particles, including liquid droplets. Combustion conditions can affect the number, particle size and chemical speciation of the particles.

Polynuclear aromatic hydrocarbons (PAH) concentrations are usually low indoors. PAH concerns stem from their potential to act synergistically, antagonistically or in an additive fashion with other contaminants. The chemical composition and concentrations of these compounds vary with combustion conditions.

Sources. Combustion products are released under conditions where incomplete combustion can occur, including: wood, gas and coal stoves; unvented kerosene heaters; fireplaces under downdraft conditions; and environmental tobacco smoke. Vehicle exhaust is a primary source, particularly from underground or attached garages.

Health Effects. The impact on human health varies with the category of combustion product; so they are treated separately below.

Carbon monoxide (CO) has about 250 times the affinity for hemoglobin than oxygen has. When carboxyhemoglobin (COHb) is formed, it reduces the hemoglobin available to carry oxygen to body tissues. CO, therefore, acts as an asphyxiating agent. Common symptoms are dizziness, dull headache, nausea, ringing in the ears and pounding of the heart. Should CO inhalation induce unconsciousness, damage to the central nervous system, the brain and the circulatory system may occur. Acute exposure can be fatal. Young children and persons with asthma, anemia, heart and hypermetabolic diseases are more susceptible.

The extent to which nitrogen oxides (NO_x) affect human health is unclear. The most information is available about nitrogen dioxide (NO_2). NO_2 symptoms are irritation to eyes, nose and throat, respiratory infections and some lung impairment. Altered lung function and acute respiratory symptoms and illness have been observed in controlled human exposure studies and in epidemiological studies of homes using gas stoves. Studies in the United States and Britain have found that children exposed to elevated levels of NO_2 have twice the incidence of respiratory illness as children not exposed.

Combustion particulates can affect lung function. The smaller respirable particles (µm) present a greater risk as they are taken deeper into the lungs. Particles may serve as carriers of contaminants, such as PAH, or as mechanical irritants that interact with chemical contaminants. Respirable particulates, as a contaminant group, are discussed later in the chapter.

The health effects of polynuclear aromatic hydrocarbons (PAH) are very difficult to determine or predict. PAH's propensity to act in concert with other contaminants complicates any effort to attribute singular cause and effect. It is known that some PAHs are carcinogens while others exhibit pro- or co-carcinogenic potential.

ENVIRONMENTAL TOBACCO SMOKE

Environmental tobacco smoke (ETS) comes from the sidestream smoke emitted from the burning end of cigarettes, cigars and pipes and secondhand smoke exhaled by smokers. Breathing in ETS is generally referred to as passive, or involuntary, smoking.

ETS contains a mixture of irritating gases and carcinogenic tar particles. Because tobacco doesn't burn completely, other contaminants are given off, including sulfur dioxide, ammonia, nitrogen oxides, vinyl chloride, hydrogen cyanide, formaldehyde, radionuclides, benzene and arsenic.

Source. A lighted cigarette gives off approximately 4,700 chemical compounds. The EPA has estimated that 467,000 tons of tobacco are burned indoors each year.

Benzene, a known human carcinogen, is emitted from synthetic fibers, plastics and some cleaning solutions; however, the most important exposure is from cigarettes. Benzene levels have been found to be 30 to 50 percent higher in homes with smokers than in nonsmoking households.

Carbon monoxide, nicotine and tar particles have been identified as the chemicals most apt to impact on health.

Symptoms and health effects. In 1985, three federal bodies independently arrived at the same conclusion: passive smoking significantly increases the risk of lung cancer in adults. The Surgeon General warned: "A substantial number of the lung cancer deaths that occur among nonsmokers can be attributed to involuntary smoking."

According to the EPA, tobacco smoke contains 43 carcinogenic compounds. The agency also notes that ETS is a major source of mutagenic substance; i.e., compounds that cause permanent changes in the genetic material of cells.

Studies have shown that passive smoking significantly increases respiratory illness in children. Asthmatic children are particularly at risk.

The federal Interagency Task Force on Environmental Cancer, Heart and Lung Disease Workshop on ETS concluded that the effects of ETS on the heart may be an even greater concern than its effect on the lungs. Several studies have linked passive smoking with heart disease.

FORMALDEHYDE

Formaldehyde (HCHO) is a volatile organic compound. It is used in a wide variety of products and is most frequently introduced into the building during initial construction or renovation. It is a colorless gas at room temperature and has a pungent odor. Formaldehyde is a ubiquitous chemical influenced by temperature and humidity.

Source. Formaldehyde is used in many building products, including plywood, paneling, particleboard, fiberboard, urea-formaldehyde foam insulation, adhesives, fiberglass and wallboard. Among potential sources are furniture, shelving, partitions, ceiling tiles, wall coverings, draperies, upholstery, carpet backing and ceiling tiles. Concentrations tend to be highest in prefabricated homes, ranging from 0.1 to 5.0 mg/m^3 (1 mg/m^3 = 0.8 13 ppm HCHO). More commonly the range is 0.1 to 1.0 mg/m^3.

Formaldehyde is also a product of incomplete combustion; therefore, cigarette smoke, and cooking/heating fuels, such as natural gas and kerosene, are sources.

Symptoms and health effects. Clinical and epidemiological data indicate human response to formaldehyde can vary greatly. Some people exhibit hypersensitive reactions. Acute exposure to formaldehyde can result in eye, ear, nose and throat irritation, coughing and wheezing, fatigue, skin rash and severe allergic reactions. It is a highly reactive chemical that combines with protein and can cause allergic contact dermatitis. Table 4-4 shows the effect of short term exposure with ranges and median responses, excluding immunosensitive populations.

Controversy remains regarding the carcinogenic role of formaldehyde in humans. The incidence of cancer in rodents makes formaldehyde highly suspect as a human carcinogen. EPA has recently conducted research which suggests formaldehyde may cause a rare form of throat cancer in long-term occupants of mobile homes. Chamber studies have shown that a given concentration of formaldehyde may evoke quite different degrees of irritation, depending upon duration of exposure, fluctuations in concentrations and the presence of other agents in the air.

TABLE 4-4.
EFFECT OF FORMALDEHYDE IN HUMANS AFTER SHORT-TERM RESPIRATORY EXPOSURE

REPORTED RANGES*	ESTIMATED MEDIAN*	EFFECT
0.06-1.2	0.1	Odor threshold 50% of people
0.01-1.9	0.5	Eye irritation threshold
0.1-3.1	0.6	Throat irritation threshold
2.5-3.7	3.1	Biting sensation in nose, eye
5-6.7	5.6	Tearing eyes, long term lung effects
12-25	17.8	Tolerable for 30 minutes with strong flow of tears lasting 1 hour
37-60	37.5	Inflammation of lung (pneumonitis), edema, respiratory distress: danger to life
60-125	—	Death

* Concentrations in mg/m^3; 1 mg/m^3=0.813 ppm
Source: National Center for Toxicological Research (1984)

RADON

Radon is an odorless, colorless gas that is always present at various concentrations in the air. Radon is formed from the decay of radium, which in turn results from the decay of uranium. Radon (Rn-222) and more specifically, the radioactive elements to which it decays (radon daughters) represent a major indoor air concern, particularly in homes.

Radon daughters are charged particles and they, in turn, adsorb or attach to particles in the air. Approximately 90 percent of the radon daughters attach to airborne particles before they are inhaled. The remaining 10 percent represent a significant source of exposure; for, as smaller particles, they deliver a dose to critical lung cells. The following section on respirable particulates discusses this aspect further. About 30 percent of the inhaled daughters are deposited in the lung.

Radon has a radiological half-life of 3.8 days, while the daughters' half-life is about 30 minutes. This rapid decay means the daughters emit high radiation energy levels to a comparatively small volume of tissue; and, in the process, provide a major source of injury to tissues.

Radon exhibits daily and seasonal variation in concentration. Fleming and others have found extreme variability, 7.5 to 25 picoCuries per liter (pCi/L) in a diurnal cycles in many buildings that do not coincide with ventilation rates. Indoor radon concentrations in most climates are much lower in the summer. This is usually attributed to more open windows and doors, which increases air flow and tends to equalize pressure differentials.

Source. Some radon will enter a facility through the water system and off-gassing of building materials; however, the principal source of radon is the soil. Radon typically enters through cracks, voids or other openings in the foundation.

Conditions affecting the flow of radon into a facility are:

soil factors - level of radon concentration, emanation rate, diffusion length, permeability, soil moisture;

building factors - type and formation of the foundation or substructure, construction quality, design; and

pressure differentials - building depressurization through stack (thermal) effect and/or exhaust fans, wind, barometric pressure changes, pressure gradients in the soil.

Symptoms and health effects. No immediate symptoms are associated with radon exposure.

When absorbed into the lung cavity, radon decay products may increase the incidence of lung cancer. There is evidence that tobacco smokers exposed to radon are more likely to get lung cancer.

RESPIRABLE PARTICULATES

Particulates represent a broad class of chemical and physical contaminants found in the air as discrete particles. Respirable particulates are generally defined as 10 microns or less in size. Particles fall into two categories; biological and non-biological.

Size determines the magnitude of risk and the ultimate location in the lungs. Since smaller particles are breathed deeper into the lungs, they can by-pass respiratory defense mechanisms thereby creating a greater health hazard. Smaller particles also stay suspended longer and, therefore, offer a greater possibility of inhalation. Smaller particles also offer a greater surface area to total mass ratio. Since respirable particles also serve as the carriers for other contaminants, such as pesticides, PCBs, radon daughters and pathogens, they can deliver harmful substances to more vulnerable areas.

Particles associated with particular contaminants, such as asbestos and bioaerosols, are treated in separate sections.

Sources. Biological particles include microbial particles, which may emit harmful organic gases in the air. Plant and animal material also supply biological particles. Common sources of respirable particulates are environmental tobacco smoke, kerosene heaters, humidifiers, wood stoves and fireplaces. Non-biological particles, including dust and dirt, are brought in on occupant clothing, come from occupant activities, maintenance products and the natural deterioration of building products and furnishings.

Symptoms and health effects. Irritation and infections in the respiratory track and eye irritation are all symptoms associated with respirable particles. All symptoms and health effects associated with environmental tobacco smoke also apply. Respirable particulates are also associated with lung cancer.

In 1986 Amman et al. listed the following concerns related to respirable particles:

(1) chemical or mechanical irritation of tissues, including nerve endings at the site of deposition,

(2) impairment of respiratory mechanics,

(3) aggravation of existing respiratory or cardiovascular diseases,

(4) reduction in particle clearance and other host defense mechanisms,

(5) impact on host immune system,

(6) morphologic changes of lung tissues, and

(7) carcinogenesis.

Health consequences vary with the size, mass, concentration and other contaminants acting in concert with the particles. EPA has found that respirable particles at concentrations of 250 to 350 $\mu g/m^3$ increase respiratory symptoms in compromised individuals.

Correlations have been found between airborne man-made mineral fibers (MMMF) and eye-irritation and between skin irritation and nonrespirable MMMF. The World Health Organization issued a report, Biological Effects of Man-Made Mineral Fibers (No. 81), in 1983.

Because of their adsorption properties, particles carry semivolatile chemicals, such as pesticides, dioxins, PCBs into humans as they inhale or ingest the particles. Health effects normally associated with these chemicals, including cancer, can be attributed to respirable particles as well.

VOLATILE ORGANIC COMPOUNDS (VOCS)

Organic compounds that exist as a gas, or can easily off-gas under normal room temperatures and relative humidity, are considered volatile. A range of VOCs are always present in indoor air.

NOTE: Formaldehyde is a VOC. Tobacco smoke and other combustion contaminants emit VOCs such as formaldehyde, benzene, phenols. For more specific treatment of formaldehyde, ETS or combustion contaminants, please see other sections in this chapter and Appendix A.

Sources. Hundreds of VOCs are found in the indoor air. The list of potential sources is lengthy. Some of the major, and more common sources, are photocopying materials, paints, gasoline, people, refrigerants, personal hygiene and cosmetic products, building materials, molded plastic containers, disinfectants, cleaning products and environmental tobacco smoke. Some of these sources emit several VOCs. For example, tobacco smoke contains such VOCs as alcohols, acetone, benzene, formaldehyde, phenols, ammonia, aromatic hydrocarbons and toluene. Few are unique to any one source, as toluene, for instance, can also be found in gasoline, paints, adhesives and solvents.

Symptoms and health effects. Symptoms attributable to VOCs include respiratory distress, sore throat, eye irritation, nausea, drowsiness, fatigue, headaches and general malaise.

Specific VOCs are not often proven to cause SBS complaints. Due to the large numbers of chemicals found indoors, it is very difficult to establish any causal relationship between health and certain VOCs. Industrial exposure studies have documented respiratory ailments, heart disease, allergic reactions, mutagenicity and cancer to some VOCs. Combinations of certain VOCs are suspected of having synergistic effects and this potential is currently being researched.

ENVIRONMENTAL CONDITIONS

Contaminants may not act alone in affecting occupant health. General environmental conditions, such as temperature and lighting, can interact with contaminants. Environmental conditions may act independently, physically, antagonistically or synergistically with various contaminants. For example, room temperature and relative humidity (RH) have a significant impact on the rate formaldehyde off-gases. RH also dictates the breeding opportunities for various agents. Air that is too humid, for example, may foster the growth of mold.

Factors related to thermal environment conditions are discussed in detail in Chapter 8. Other factors that may serve as a source of IAQ symptoms or interact with the contaminants include artificial light, noise and vibrations, particles and fibers, ions, psychological and ergonomic considerations.

ARTIFICIAL LIGHT

Poor lighting conditions can cause eye strain and irritation as well as headaches and may increase sensitivity to certain contaminants. Visual stress may come from insufficient contrast in the material, brightness, glare and inappropriate light levels.

Brightness

Brightness is determined by the relative amount of light available at the work surface in relation to the level of illumination in the field of view. The eyes functions most comfortably and efficiently when the brightness relationships are not excessive. The light at the desk surface, for example, should not be more than three times the level of light immediately around the desk. A desk lamp which is being used in a dark room exceeds the recommended brightness ratio and reduces eye comfort and efficiency.

Glare

In maintaining a lighting system, conditions which create glare should be avoided. The position of the light source in relation to both the viewed surface and the eye is critical. For example, light in front of a desk that strikes the surface and is reflected into the eye can create considerable eye strain. This effect, called veiling reflection, is shown in Figure 4-2.

A simple way to check for veiling reflection is to place a mirror on the work surface in front of the worker. If the reflection from the mirror strikes the worker's face, then that person is being subjected to unnecessary glare. Light sources that cause this effect should be removed or repositioned; or, in the case of movable work surfaces, the work area should be repositioned in relation to the light.

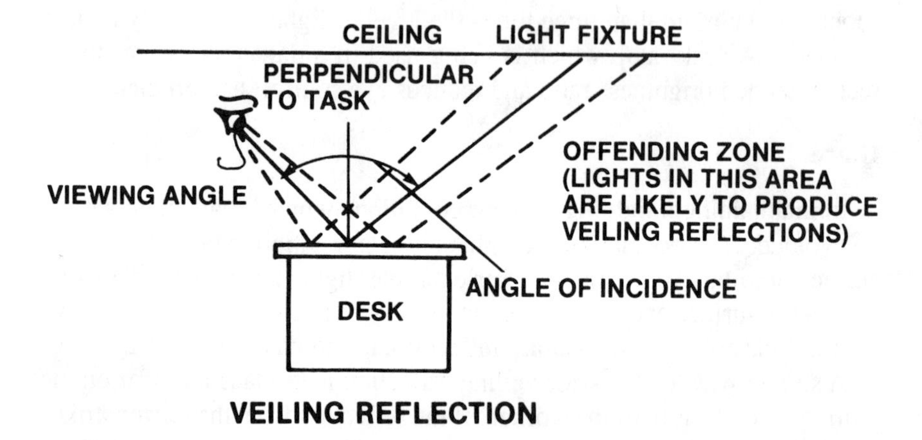

FIGURE 4-2. VEILING REFLECTION

Quantity of Light

A number of sources discuss lighting and recommended lighting levels. The Electrification Council's manual, Fundamentals of Commercial and Industrial Lighting, available through participating utilities, is an excellent source on lighting. The Illuminating Engineering Society of North America (IES) has published recommended levels of lighting in its IES Handbook, Reference Volume. The lighting section in ASHRAE 90A-1980, authored by IES, is a good reference. Lamp manufacturers also have information available. For example, Table 4-5 shows the generally recommended levels of illumination in footcandles appropriate for educational facilities.

TABLE 4-5. RECOMMENDED ILLUMINATION LEVELS FOR EDUCATIONAL FACILITIES

AREA/ ACTIVITY	FOOTCANDLES ON TASK USAGE CATEGORY*		
	I	II	III
Educational facilities			
Classrooms (general)	50	75	100
Science laboratories	50	75	100
Lecture rooms			
Audience	50	75	100
Demonstration	100	150	200
Auditoriums			
Assembly	10	15	20
Social activity	5	7.5	10
Music rooms	50	75	100
Sight savings rooms	100	150	200
Study halls	50	75	100
Typing	20	30	50
Corridors	10	15	20
Toilets and washrooms	10	15	20
Food service facilities			
Dining areas			
Cashier	20	30	50
Cleaning	10	15	20
Dining	5		10
Kitchen	50	75	100
Gyms	30		
Locker rooms	10	15	20
Offices			
Lobbies, lounges & reception	10	15	20
Offset printing & duplication area	20	30	50
General and private	20	30	50
Shops			
Woodworking	20	30	50
Rough bench work	20	30	50
Fine bench work	200	300	500
Maintenance facilities			
Garages-repairs	50	75	100

*Usage categories I, II, and III are determined by weighting factors which consider occupants ages, room surfaces reflectances, speed and/or accuracy of task involved, and reflectance of task background. See IES Handbook, Reference Volume, 1981 for complete discussion of method for determining illumination levels.

Some fluorescent lamps have had a negative affect on hyperactive children. Earlier high pressure sodium (HPS) lamps were thought to cause headaches and malaise. Wilkins et al. reduced the fluctuation in illumination using a solid state high frequency ballast and decreased eye-strain and headaches by 50 percent in a group of office workers. Recent studies suggest full-spectrum lighting reduces absenteeism. VDT users require particularly well-designed lighting.

NOISE AND VIBRATION

Noise has been found to affect human health by volume, sound pressure levels, infrasound and frequency. Noise at 70-80 decibels (dB) is suspected of causing tiredness. The Occupational Safety and Health Administration sets industrial guidelines at 90 dB. In comparing two situations where the sound levels were both approximately 60 dB and room features were similar, frequency analysis revealed sound pressure levels in the 8-125 Hz range was much higher in the area described as "unpleasant" by the workers. In situation where levels are above 120 dB, infrasound (sound waves in 0.1 - 20 Hz range) may cause dizziness and nausea. The more common situation to cause problems is found where industrial machines or ventilation equipment emit low frequency noise (20-100 Hz).

Several smaller studies have found a relationship between health and vibrations. The causative link is believed to be the characteristic resonance frequencies (1-20 Hz) of certain body parts, especially the eye, and external vibrations. A significant correlation was found between office worker irritability and dizziness and the level of vibrations (from a nearby pump room) measured at their desks.

The International Standard Organization (ISO) has done extensive work on acoustics and on the effects of vibrations on the human body. Information on their reports, Evaluation of Human Exposure to Whole-Body Vibration (1985) and Acoustics, Description and Measurement of Environmental Noise (1987), can be obtained by contacting ISO in Geneva.

IONS

With the advent of office equipment requiring high power charges, such as computers, considerable concern has been evidenced about the relationship of negative ions in the atmosphere and SBS. To date, the results are inconclusive.

Guilleman describes negative ionizers as releasing significant amounts of ozone, which is considered a potent throat irritant. In a double blind study, Finnegan et al. found the level of symptoms of SBS was not influenced by the concentration of ions in the air.

PSYCHOLOGICAL AND ERGONOMIC FACTORS

A direct causal relationship between psychological factors and SBS has not been established. The World Health Organization reported in 1986 that there is some evidence that stress, which may be the result of some psychological factors, can make individuals more susceptible to environmental factors. Work by Hedge et al. and Morris et al., while far from conclusive, suggests that SBS may be responsible for stress, rather than the reverse which many have assumed.

There is an increasing body of information regarding the health effects of computer work, particularly visual display terminals (VDTs). There is a consensus that prolonged VDT exposure can cause eye irritations, headaches, tiredness and appears to be a factor related to more general IAQ complaints. There has been some indication that the incidence of miscarriage is related to work with VDT; however, more research in this area appears warranted.

One of the most telling psychological factors related to the modern work environment may be embodied in a factor listed by McDonald in 1984. He found the fact that workers have little or no control over regulation of temperature, humidity and lighting at their work location was a common problem relative to working conditions.

The effects of ergonomics has been studied in Europe for a number of years. It has more recently emerged as a concern in the United States. Defined as the study of problems people have in adjusting to their environment or the science of adapting working conditions to the worker, it generally refers to the physical conditions. Stettman et al., reported in 1985 that they found ergonomic factors influence the perception of indoor air quality.

HEALTH, CONTAMINANTS AND ENVIRONMENTAL FACTORS

Table 4-6 summarizes the major SBS factors related to the paradigms offered above to present the relationships of health effects, contaminants and environmental factors. Table 4-7 summarizes the health effects, contaminants and sources associated with building related illnesses.

The data assembled in Tables 4-6 and 4-7 were digested from a careful review of the literature related to SBS and BRI health effects. The tables are by no means inclusive or comprehensive; they only highlight conditions related to each category. They are certainly not medically definitive. To be as accurate as possible, the terms used by the epidemiologists and others reporting their research were used verbatim; e.g., "dry cough" and "coughing," and no medical interpretations were attempted.

The material is not intended for medical diagnosis. Rather, the tables are designed to help the practitioner sort through what is known, the symptoms, and to discern possible contaminants/sources in Table 4-6 and possible BRIs/contaminants/sources in Table 4-7. An administrator using Table 4-7, for example, may discern that an occupant appears to have the symptoms in Complex A, which would suggest medical involvement is warranted, bioaerosols are suspect and investigation of the related primary sources should be considered.

If Tables 4-6 and 4-7 suggest the process is easy or straightforward, then they are misleading. It is seldom a straight shot from symptom to contaminant to source. The search, unfortunately, is usually a sorting process of exclusion. It is confused by the fact that several contaminants can manifest the same symptoms. Nitrous oxides and formaldehyde, for example, evidence strikingly similar symptoms. Some dusts and various microbial contaminants prompt similar allergic type reactions.

Despite the confusion, it is easier working from the known; i.e., symptoms, than identify contaminants and trying to find our way back. In many instances, the causes are never identified and the contaminant-to-remedy process never even gets started.

A review of the tables will reveal that they merely condense major findings offered throughout this chapter; so the practitioner can, at a glance, sense cause/effect relationships that may be at work in a given facility. To pursue particular contaminants/sources/environmental conditions as they relate to certain health effects, the reader is urged to refer to earlier portions of this chapter and Appendix A.

In using the tables, the reader should keep in mind that every health effect cited may be the result of medical conditions not associated with SBS or BRI.

TABLE 4-6. THE PRACTITIONER'S GUIDE TO HEALTH EFFECTS, CONTAMINANTS AND ENVIRONMENTAL FACTORS

SICK BUILDING SYNDROME

SYMPTOM COMPLEX / ILLNESS HEALTH EFFECT	POSSIBLE CONTAMINANTS	PRIMARY SOURCES	ENVIRONMENTAL CONDITIONS
Eye Irritation Burning, dry gritty eye without inflammation	NO_2	Incomplete Combustion - stoves, fireplaces, ETS	Artificial light
	Formaldehyde	Building products & furnishings	
Watery eyes	VOCs	Broad range of products (See VOC section)	
	Bioaerosols	Ventilation systems, humidifiers, dehumidifiers wet insulation, drip pans, cooling coils in AHUs, people, pets, plants, insects outside air	
	Particulates	Combustion products, esp. ETS, man-made fibers, dust, dirt, maintenance products, building product deterioration	
Nasal manifestations "Stuffiness" Nasal irritations, rhinorrhoea	NO_2	Incomplete combustion - stoves, fireplaces, ETS	Relative humidity High temperatures
	Formaldehyde	Building products & furnishings	
	Bioaerosols	Ventilation systems, humidifiers, dehumidifiers wet insulation, drip pans, cooling coils in AHUs, people, pets, plants, insects outside air	

TABLE 4-6. SICK BUILDING SYNDROME continued

SYMPTOM COMPLEX / ILLNESS HEALTH EFFECT	POSSIBLE CONTAMINANTS	PRIMARY SOURCES	ENVIRONMENTAL CONDITIONS
Throat, Lower Respiratory Tract	NO_2	Incomplete combustion - stoves, fireplaces, ETS	Relative humidity
Dry throat, no inflamamation Shortness of breath without lung infections or bronchial asthma	Formaldehyde	Building products & furnishings	
	VOCs	Broad product range (See VOCs section)	
Lung Cancer Irritation & infection of respiratory tract	ETS	Passive Smoking	
	Particulates	Esp. combustion products	
Headache, Fatigue, Malaise Headaches - frontal afternoon occurence Poor concentration Dizziness Tiredness Irritability	Bioaerosols	Ventilation systems, humidifiers, dehumidifiers wet insulation, drip pans, cooling coils in AHU, people, pets, plants, insects outside air	Ergonomic conditions Noise and vibrations
	VOCs	Broad range (See VOC section)	
With nausea, ringing in ears, pounding heart	CO	Incomplete combustion - vehicle exhaust - stoves, fireplaces, ETS	
Fatigue	Formaldehyde	Bulding products & furnishings	
Skin problems dryness, irritation			Warm air, Low relative humidity Excessive air movement
Rashes (improves away from building)	Formaldehyde		

TABLE 4-7. THE PRACTITIONER'S GUIDE TO HEALTH EFFECTS, CONTAMINANTS AND ENVIRONMENTAL FACTORS

BUILDING RELATED ILLNESSES

SYMPTOMS	POSSIBLE BRIs	POSSIBLE CONTAMINANTS	PRIMARY SOURCES
Rhinitis Asthma-like symptoms Other allergic reactions	Allergies	Allergens - microbial	Poor maintenance Humidifiers, esp. cold spray
		Allergens - chemical e.g., formaldehyde	Building products & furnishings
SYMPTOM COMPLEXES			
COMPLEX A - headaches* - fever & chills* - dry cough * - lung disorders - malaise & myalgia - shortness of breath (worsens during exposure)	Hypersensitivity pnuemonitis (extrinisic allergic alveolitis)	Organic dust (microorganisms, endospores, animal protein) Low molecular weight chemicals	People, pets, plants, insects Outside air Ventilation systems, humidifiers, dehumidifiers, wet insulation, drip pans, cooling coils in AHUs
COMPLEX B - fever & chills* - coughing* - lethargy - arthralgic & myalgia - polyuria - weight loss - breathlessness (lessens during exposure)	Humidifier fever	Microorganisms (incriminated organisms not yet indisputably identified)	Humidifier

TABLE 4-7. BUILDING RELATED ILLNESSES continued

SYMPTOMS	POSSIBLE BRIs	POSSIBLE CONTAMINANTS	PRIMARY SOURCES
COMPLEX C- headache* - fever & chills* - dry cough* - chest pains, shortness of breath - vomiting. abdominal pain - diarrhea - weight loss - myalgia - pnuemonia	Legionnaire's Disease (LD)	Legionella pneumophila	Aerosols from cooling towers, evaporative condensers, shower heads, water faucets and hot water systems
COMPLEX D Milder form of LD No pnuemonia	Pontiac fever	Legionella	Aerosols from cooling towers, evaporative condensers, showers water systems
COMPLEX E (Infectious) - headaches* - fever & chills* - cough* - respiratory infections - broad range of other symptoms	Fungal infections	Over 100,000 fungi have been described. Of interest are aspergillus, mycotoxins and endotoxins. Molds and spores, or saprophytic fungi (about 300,000 species)	Outside air, duct work People, pets, insects, plants Water damaged materials, organic sources
	Viral infections	Viruses	People, pets
HEALTH EFFECTS LONG TERM			
No immediate symptoms	Asbestosis Mesothelioma & lung cancer	High level exposure to asbestos fibers Primarily amphiboles asbestos, not common in U.S.	Primary industrial exposure Asbestos Containing Materials (ACM)
	Lung cancer	Radon, radon daughters Particulates	Soil (water, building materials) Combustion product

*Symptoms common to all BRIs

Chapter 5

Investigating Indoor Air Problems

When indoor air problems mushroom into full scale concerns, owners and operators have a tendency to turn immediately to outside expertise. A lot can be done, however, through simple in-house steps before such a move becomes necessary. In fact, a walk through survey and a few corrective actions may avoid the call for outside consultation entirely. Should it become necessary to engage an IAQ diagnostic team, the preliminary work by the staff can greatly expedite the team's efforts.

When a building is "sick," two steps can be taken before the "doctor" arrives. Staying with the medical analogy, the first step is to "Take two aspirin and call me in the morning;" i.e., do the simple things in the hope that the doctor will not be needed in the morning. The second step is "What to do until the doctor comes."

If the "doctor" is needed, treatment may be a simple prescription, or the doctor may find that some extensive tests are required. In either case, some cataloging of symptoms, their patterns and building conditions can prove helpful.

If symptoms appear to be serious, life-threatening, or are likely to cause long-term health damages, a diagnostic team should be brought in immediately and the investigations should proceed as rapidly as possible. Immediate steps should be taken to protect the occupants in such a situation, including evacuation of the area or the building as appropriate.

It should be stressed that the investigation process does not replace a disaster plan, such as procedures to respond to a chemical spill in or near a facility. Nor does it address long term hazards.

From the simple inspection to complex diagnostics and post- treatment monitoring, IAQ building investigations can be divided into the five phases:

(1) Preliminary Assessment -- a self-evaluation, data gathering, observation effort;

(2) The Walk Through Inspection -- conducted by trained in-house staff, or as a preliminary inspection by the diagnostic team. Measurements are generally confined to the use of a smoke pencil and gauges for temperature and humidity. 80 percent of IAQ problems are usually identified by this stage of the investigation process;

(3) Simple Diagnostics -- more extensive analytical procedures conducted by the diagnostic team; limited measurements of implicated factors or surrogates; possible medical assessment;

(4) Complex Diagnostics -- broad in-depth testing; qualitative studies of factors in combination; medical examinations; and

(5) Monitoring and Recurrence Prevention -- observation, testing as warranted; preventive measures.

With a little background information and training, the first phase can easily be handled internally. Depending on the staff's level of expertise, in-house personnel may conduct part, or all, of the second phase (the walk through inspection). Even with trained staff, however, it may be advisable to get outside consultation during this phase should the emotional climate or the seriousness of the IAQ problem warrant it. Phases 3 and 4, Simple and Complex Diagnostics, require the special expertise of a multidisciplinary team.

Management concerns related to problem investigation focus on in-house investigation strategies associated with Phases 1 and 2. If the diagnostic phases become necessary, management responsibilities then extend to data preparation and consultant selection, investigation support and oversight. In order to appreciate the work that will be required of a diagnostic team, management needs a general understanding of diagnostic procedures. The more detailed technical and contaminant-specific procedures do not generally fall within the management purview and are not treated in this text.

Each phase of an investigation is apt to suggest remedies. Before proceeding to the next phase, corrective measures should be taken and the effects of those measures observed. The move to the succeeding phase(s) will depend on how well the preceding efforts have worked in alleviating the problem(s). The progression of the investigative phases, appropriate personnel related actions and evaluation procedures are shown in Figure 5-1. Note that each phase is followed by monitoring procedures and a decision as to whether or not the investigation needs to progress to the next phase.

INVESTIGATION PROCEDURES

As with any new field, there seems to be an abundance of people, who are prone to embroider and embellish IAQ concerns with Latin phrases and technical jargon. Others get caught up in their fields of expertise.

Medical confirmation supplied a couple months later may be interesting, but hardly vital to the building manager who has the problem now.

It is not necessary to know such technical information as "Q fever is a zoonosis caused by Coxiella burnetti," to get to work. Scientists want to know specific causes; managers want solutions.

WHERE TO START

The best investigation procedure currently available is the solution-oriented approach. Since there is so much we don't know about detecting indoor air problems and so many difficulties associated with measurement and verification procedures, the simplest process is to start with what we do know and chip away at it.

With the knowledge afforded from other sections of this book, it may be possible to spot the problem almost immediately and fix it. If that is the case, a place on the complaint form (Chapter 3) can be used to indicate what was done, when it was done, who did it and any follow-up monitoring that is needed.

If, however, the cause is not immediately obvious, or a broader investigation seems appropriate, a preliminary assessment of the situation should be the first step.

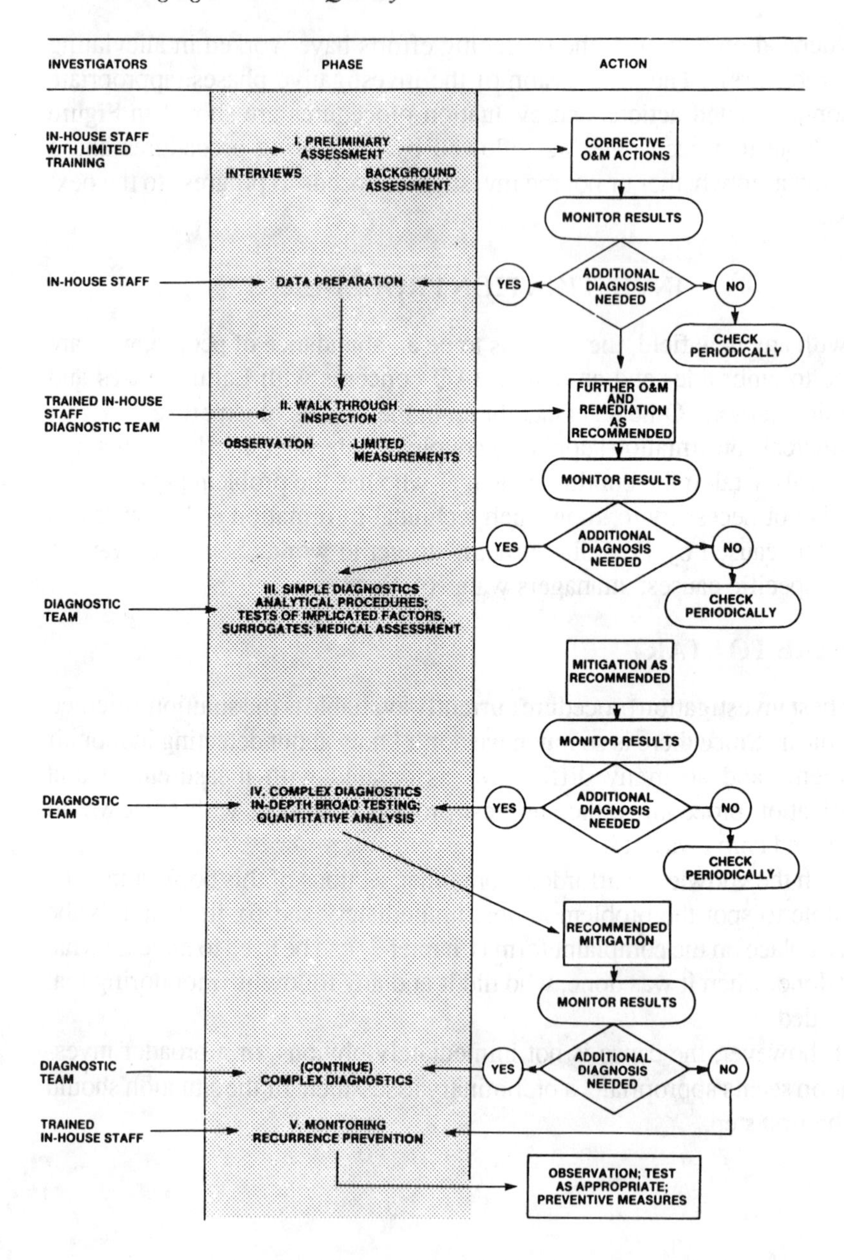

FIGURE 5-1. INVESTIGATIVE PHASES

A prompt site visit not only aids in identifying many sources, it can reduce the spread of the illness, contain economic losses and demonstrate a caring attitude toward the affected people. Delays aggravate inspection difficulties as physical changes are frequently compounded over time. Irritated personnel, who feel their concerns have been ignored, can cloud or exaggerate the situation. Early recognition of a problem as well as timely and systematic evaluation are key factors to a quick and effective resolution.

PHASE I: PRELIMINARY ASSESSMENT

The Preliminary Assessment is designed as a first response to reports of indoor air problems. It is a data gathering and observation effort. No measurements are taken. No professional consultation is required.

The purpose of the Preliminary Assessment is two fold: (1) to get a better understanding of the extent of the problem and related conditions; and (2) to identify possible causes. The assessment procedure has two distinct facets, determining: (a) the nature and scope of the complaints; and (b) a preliminary audit of facility conditions, systems, maintenance and operational procedures.

The observations are best done by a team. Someone on the team should have knowledge of HVAC system operations. It is valuable to have someone from outside the facility (not necessarily outside the organization), who can take a fresh look at conditions. An effective preliminary assessment process does require that those conducting the assessment have some knowledge of indoor air quality concerns and acceptable data gathering procedures. The training need not be extensive, but effective self-evaluation needs some guidance.

The areas examined by the team should be influenced by the nature and locality of the complaints. If an occupant has been medically diagnosed as having hypersensitivity pneumonitis, the contaminant is a bioaerosol and testing for formaldehyde would not be the first order of business. This is not to say that a broader investigation may not be warranted if problems persist, but the order of the day is to find the source of the identified problem.

Malfunctions in the ventilation system may have allowed concentrations of a contaminant to reach the level that affects health. During the Preliminary Assessment, a visual check of the system should be made to be sure it is performing satisfactorily.

Caution is warranted: The Preliminary Assessment phase is not intended to, nor will it, resolve many indoor air problems. It is as it's name describes, a preliminary effort. Neither are the procedures designed to treat indoor air problems that are not manifested in SBS symptoms, such as asbestos and radon. These contaminants must be treated by inspection and control procedures discussed in Chapter 6 and Appendix A.

Nature and Scope of the Complaints

Determining the extent of the problem will usually involve some preliminary interviews to discover the area(s) in which the symptoms are evidenced, the pattern, the duration and the number of people experiencing similar symptoms or medical signs. Psychological and ergonomic factors may also need to be considered. Some of this information may be available through a review of IAQ complaint forms.

Symptom information an interviewer might ask of complainants would include:

How long have you worked in the area?

When did you first notice the symptom?

What symptoms have you experienced? (headaches, chest pains, shortness of breath, nose and/or throat irritation, nausea, flu-like symptoms, abdominal pains, drowsiness, lethargy, fatigue, fever, eye irritation, skin rash or itching, etc.)

Do you suffer from any medical problems that may be related to these symptoms? (asthma, allergies, hay fever, eczema, migraines, arthritis)

Do you take any medication? (over the counter or prescribed)

How often do the symptoms occur? (___ times per day, week, month)

Do the symptoms occur during a particular part of the year? (winter) Week? (Monday morning) Or time of day? (First thing in the morning, end of the day)

How long do they last? (all morning, all day, all week, all the time)

Are there any particular areas in which you work where the problem seems greater?

Are there any specific tasks you perform just before symptoms are noted?

Do symptoms go away upon leaving the building?

Do you wear contact lenses? Operate visual display terminals or copiers more than 10 percent of the day? Operate any special equipment? Do you work extensively with carbonless copy paper?

Do you smoke? Are you bothered by others smoking in your work area?

How would you describe the environment where you work? (noticeable odors, too hot, too cold, stuffy, dusty, noisy, etc.)

An interview form, which addresses these matters, is presented in Appendix C.

To the extent that the symptoms suggest a probable source, the facility inspection may then focus on that source. For example, in classifying IAQ problems, a very distinctive "Monday Morning Phenomena" was discussed, which is associated with humidifier fever. If the interviews reveal this symptomatology during the heating season, then at the top of the facility inspection list should be humidifiers.

Interviews and Questionnaires

Talking to the people who have registered complaints and those who have not is a valuable means of securing information about the nature and scope of the problem. Non-complainants can help pinpoint odor or temperature problems. They can also help establish that some symptoms may not be building generated. Care should be taken in the interview process to obtain objective data and avoid biasing the data by suggestion.

At present, there is no consensus as to the value of a questionnaire in the preliminary assessment process. To some degree, this can be attributed to the perspectives of the disciplines involved. Some think questionnaires are indispensable and others have no use for them at all. Just a few thoughts from respected IAQ authorities, will offer a glimpse of the dichotomy. One source says,

> Questionnaires have been shown to be powerful tools in screening and clinical diagnosis. They also play an important role in indoor air quality investigations.

The National Institute of Occupational Safety and Health (NIOSH) has found that when the questionnaire is administered prior to the initial visit, the findings enable the NIOSH team to develop more effective strategies in dealing with the problem and to use of the investigators' time on site more efficiently.

On the other hand, Charles Lane of Honeywell Indoor Air Quality Diagnostics (IAQD) has pointed out that while questionnaires may be valuable tools for epidemiologists to establish gross averages, they are easily misapplied in individual situations. Lane observed that getting data by bits doesn't tell a diagnostic team anything. The Honeywell IAQD team does on occasion use a one-page questionnaire during on-site investigations to determine if real time correlations exist between subjective and objective data.

Gray Robertson of Healthy Buildings International, Inc. (HBI) a firm with extensive SBS investigation experience, states emphatically,

> Furthermore, due to their unreliability, we, as a policy, refuse to rely upon or otherwise use the information generated by subjective building occupant questionnaires.

The questionnaire is a specialized tool for IAQ investigators that can easily be misused by lay persons in obtaining and interpreting IAQ data. If the investigative process is to benefit from the use of questionnaires, they must be administered and interpreted by those skilled in their use. For this reason, only an interview form has been provided in Appendix C.

When used correctly, the interview procedure should help define the complaints, assist in determining if the problem is localized and whether any special circumstances; e.g., activities, time of day or week, improves or worsens the problem.

Background Assessment

Prior to any extensive background assessment, it is best to ascertain whether operational conditions in the affected area are abnormal. Or, were abnormal just prior to the IAQ episode. If complaints have been generated by an abnormality, rectifying the situation may satisfy the problem and no further assessment may be needed.

Background assessment procedures seek to identify what is, and what ought to be. The basic purpose is to assess the overall condition of the building, what changes have been made and whether systems are operating properly. Original construction information, such as construction dates, square footage, ventilation system design, materials used, etc., is gathered along with any changes that could affect thermal or contaminant loads. A review of changes and modifications helps to identify the possible introduction of contaminants from building materials and furnishings as well as unmet ventilation needs. Inspection procedures should note any changes in space utilization, such as relocation of copiers and laser printers to poorly ventilated areas of the office. The background assessment procedures can also serve a valuable function in verifying the information collected in the interviews.

As information regarding the original design, construction, subsequent changes and use are assembled, the assessment also seeks to identify probable contaminant sources. Years of IAQ investigation has given us a pretty good idea of where those sources are. During the assessment process, an "educated eye," which is sensitive to probable sources, can recognize many opportunities to improve indoor air quality.

Early Detection of Possible Resources

NIOSH, after more than 500 IAQ building investigations, has settled on a solution-oriented approach. The approach is one of exclusion. They progressively eliminate the most likely causes and gradually narrow the range of possible sources. The NIOSH approach is reminiscent of the sculptor, who explained the best way to carve an elephant was to take a piece of stone and chip away everything that doesn't look like an elephant.

A solution-oriented approach can effectively guide the initial steps of the lay person seeking solutions in his or her own building. Many problems could be averted or resolved in-house if checks were made of the areas where the problems are most likely to exist.

could be averted or resolved in-house if checks were made of the areas where the problems are most likely to exist.

A checklist of the typical sources for contaminants can be compiled from the primary types of problems found by major investigating teams. Even though some IAQ episodes have been found to be multifactorial, for our purposes they can still be classified by the primary type of problem found. A review of the frequent IAQ problems found by NIOSH, Honeywell and HBI shown in Table 5-1 reveals a commonality that can serve as a sound basis for early detection of probable causes.

TABLE 5-1. SICK BUILDING SYNDROME PROBLEMS

Org.	NIOSH	HONEYWELL	HBI
Bldgs.	529	50	223
Yr	1987	1989	1989
	Inadequate ventilation (52%) Inside contamination (17%)	Operations & Maintenance (75%) – energy mgmt. – maintenance – changed loads	Poor ventilation – no fresh air (35%) – inadequate fresh air (64%) – distribution (46%)
	Outside contamination (11%) Microbiological contamination (5%)	Design – ventilation/ distribution (75%) – filtration (65%) – accessibility/ drainage (60%)	Poor filtration – low filter efficiency (57%) – poor design (44%) – poor installation (13%)
	Building fabric contamination (3%)	Contaminants (60%) – chemical – thermal – biological	Contaminated systems – excessively dirty duct work (38%) – condensate trays (63%) – humidifiers (16%)

Comparing these lists becomes a process of using what these organizations have learned to your advantage. Ventilation, as a source of indoor air problems, is shown on every list. It makes good sense, therefore, to carefully assess the conditions of the HVAC system, as the building assessment is conducted.

Table 5-1 also reveals that HBI and Honeywell have found ventilation distribution frequently shows up as a problem. NIOSH's category labels are not as informative; however, if the NIOSH findings are broken down further, distribution is listed as a major contributor to inadequate ventilation.

This observation serves to point out that there is a wealth of information to be drawn from a closer look at the NIOSH investigation results. This information can highlight areas or conditions staff investigators should be looking for.

Inadequate Ventilation

- Not enough fresh outdoor air supplied to the office space;
- Poor air distribution and mixing which causes stratification, draftiness, and pressure differences between office spaces;
- Temperature and humidity extremes or fluctuations (sometimes caused by poor air distribution or faulty thermostats);
- Improper or no maintenance to the building ventilation system; and
- Energy conservation measures: reducing infiltration and exfiltration; lowering thermostats or economizer cycles in winter, raising them in summer; eliminating humidification or dehumidification systems; and early afternoon shut-down and late morning start-up of the ventilation system.

Inside Contamination

- Copying machines: methyl alcohol from spirit duplicators, butyl methacrylate from signature machines, ammonia and acetic acid from blueprint copiers;
- Improperly applied pesticides, such as chlordane;
- Boiler additives, such as dimethyl ethanolamine, which can cause dermatitis;

- Improperly applied cleaning agents, such as rug shampoo;
- Tobacco smoke of all types;
- Combustion gases from cafeterias and laboratories; and
- Cross-contamination from poorly ventilated sources leaking into other air handling zones.

Outside Contamination

- Motor vehicle exhaust, particularly from parking garages;
- Boiler gases;
- Re-entrainment of previously exhausted air caused by improperly located exhaust and intake vents or periodic changes in wind conditions;
- Construction or renovation projects including asphalt, solvents, and dusts; and
- Gasoline fumes infiltrating basements and/or sewage systems from ruptured underground tanks.

Microbiological Contamination

- Contamination in the ventilation system from bacteria, fungi, protozoa and microbial products; and
- Microbiological contamination commonly resulting from water damage to carpets or furnishings, or standing water in ventilation system components.

Building Fabric Contamination

- Formaldehyde off-gassing from urea-formaldehyde foam insulation, particle board, plywood, and some glues and adhesives commonly used during construction;
- Building materials and products;
- Fibrous glass erosion in lined ventilating ducts, causing dermatitis;
- Organic solvents from glues and adhesives; and
- Acetic acid used as a curing agent in silicone caulking.

Findings from NIOSH as cited above as well as those from other investigators can be converted to questions to guide in-house inspections, including:

> Is the air flow restricted? Are the diffusers open and unobstructed? Are they adjusted to avoid drafts? Is the exhaust system operating correctly? Are the air intakes unobstructed and operating correctly? Are air intakes bringing the building's own exhaust back into the facility?
>
> Are the filters accessible and clean? Do the filters fit the opening?
>
> Are the coils and duct work clean? Are make-up dampers functioning properly during occupied hours? Are belts and baffles functioning properly?
>
> Are rust inhibitors; i.e., volatile amines, getting into the steam? Is there standing water from the humidifiers? Or is humidifier moisture getting into nearby duct work?
>
> Has the office layout changed so that the ventilation design no longer meets occupant needs?
>
> Have inappropriate energy conservation measures, such as plastic taped over air intakes, increased indoor air problems?
>
> Do drain pans have the proper inclination that allows for continuous drainage?
>
> Is damp insulation providing a breeding ground for microbial contaminants?

Such questions need to be asked if in-house staff is to find sources of indoor air pollution rooted in poor maintenance and improperly operating equipment. To help guide such an investigation, these questions and others have been incorporated in the form presented in Appendix C. The Preliminary Assessment Form provides space for a summary of the interview information as well as building audit data.

Using the Preliminary Assessment Form

The forms should be used when the cause of the problem is not immediately apparent to the investigator. They are designed to help in-house personnel immediately investigate and solve indoor air quality problems with their own resources wherever possible. They can, of course, be modified to fit local needs.

Those who are familiar with the health effects-source-control relationship will find the forms more useful and are more apt to spot problems as they gather the data. Sharing appropriate sections of this book with the personnel conducting the assessment will be of benefit to them. The chapter on HVAC systems will also help in-house investigators become more sensitive to problems which may originate in the system.

The forms are not intended to guide in-depth inspections. Rather, they are designed to assure that the preliminary review does not miss some vital factor that in-house staff could treat. They will also help assemble the data that may be needed later by outside consultants.

At the end of Phase 1, most in-house teams will find some HVAC or general maintenance items that are contributing to the problem. They are maintenance items that should be done anyway.

While the situation cannot help but benefit from rectifying these maintenance oversights, the IAQ problems may persist. While the natives get progressively restless and irritated, there is a great temptation to go beyond the "Take two aspirin..." and start "tweaking" the system. Trying to solve one problem, the uninformed, even with the best of intentions, are apt to create others. Diagnostic teams report that they repeatedly find "a maze of maintenance misapplication" that comes from these good intentions. Sound routine maintenance is warranted, but tweaking should be resisted.

Data Preparation

Should IAQ problems still persist "in the morning," then a more extensive examination is warranted. Unless the in-house staff has achieved "paramedical" stature, it is time to shift into the "What to do until the doctor comes" mode of operation. At this point, the data on the Preliminary Assessment Form should be verified. Any maintenance measures taken in response to the Phase 1 investigation as well as any observed results should be noted.

The selected diagnostic team may request additional advance data depending on the nature of the problem and a review of the preliminary findings.

PHASE 2. THE WALK THROUGH INSPECTION

The Walk Through (WT) Inspection may be conducted by in-house personnel if they have had the requisite training. Unfortunately, the limited availability of IAQ training for facilities managers, maintenance engineers, etc. continues to constrain the staff's ability to meet their own needs. More frequently, the WT Inspection is conducted as the initial activity of a diagnostic team.

Even if the staff is adequately trained, the emotional climate and the perceived seriousness of the IAQ problem may warrant the immediate support of outside consultants.

In contrast to the Preliminary Assessment, the WT Inspection involves a more thorough examination of the facility and the HVAC system. It usually involves the use of temperature and humidity gauges and a smoke pencil to check air flow. CO_2 measurements are sometimes taken to assess contaminant concentration levels and the effectiveness of the ventilation system.

Depending on the nature and scope of the symptoms exhibited and the initial facility findings, inspectors may pursue specific concerns in greater depth. WT Inspections frequently check the same areas and equipment considered in the Preliminary Assessment, but back it up with measurements and more expertise. WTs are designed to verify Preliminary Assessment findings and address such concerns as:

Changes in building use; e.g., retail to office space, hospital to doctors' offices, and their implications for indoor air quality.

Effectiveness of the ventilation system. Deterioration of return and supply paths.

Changed thermal and contaminant loads due to new furnishings, equipment or increased occupancy.

Inappropriate energy management measures, such as closed supply louvers and air handling units turned off during the day or within certain temperature ranges. Possible modifications to VAV systems that could improve indoor air quality.

Filter efficiency and performance. Static pressure gauges as indicators. The need for additional filters downstream from contaminated heat exchange components; i.e., plenums and duct work.

Types of humidifiers (recirculating or independent steam humidification and condition; filter plate type) general cleaning schedules. Air humidification levels, especially in winter to meet the recommended relative humidity of 20-50 percent.

Measurements of the degree of circulation. Measurements of supply and exhaust flows. Ventilation systems out of balance.

Level of custodial care, maintenance; access to equipment.

Inadequate temperature controls, including inadequate recovery time after night or week-end set back.

Cleaning procedures (carpets) that leave irritating residues.

Non-porous surfaces where microbial growth is evidenced, such as cooling coils and drain pans.

Appropriate procedures for cleaning, disinfecting or using proprietary biocides -- making sure that cleaners, especially biocides, are removed before AHUs are reactivated.

Job pressures, ergonomics, management-labor stress and ways these pressures may exacerbate work area conditions that would otherwise be tolerable.

PHASE 2 SAMPLING TECHNIQUES AND ASSESSMENT PROCEDURES

Temperature and Humidity

Ambient or operative temperatures should be checked. The ASHRAE published guidelines are intended to achieve thermal conditions, which are "comfortable" for 80 percent of the occupants in a given environment. ASHRAE Standard 55-1981, Thermal Environmental Conditions for Human Occupancy, temperature recommendations consider the level of occupant activity, clothing, relative humidity (RH). For example, 80 percent of the occupants in a typical office in summer with 50 percent RH should be comfortable in the range of 73° to 79° F. If the operating temperature is outside this range, then more than 20 percent of the healthy people are apt to experience some discomfort. Similarly, winter readings at 30 percent RH should fall between 68.5° to 75° F.

When feasible, temperature checks should be taken during the conditioning season; i.e., for the heating season while the boiler is on and the system is fully operational. Efforts should be made to avoid days with extremely high or low seasonal variations.

Local thermal discomfort should be checked. The vertical temperature difference should not be greater than 5° F. from head to toe, approximately 67 inches to 4 inches.

Relative humidity (RH) has long been a concern in very humid climates. More recently, the health implications for low humidity levels have received greater attention. RH below 20 percent is now associated with increased discomfort, drying of the mucous membranes and the frequency of colds. Excessive humidity, above 60 percent, can foster contamination. (See the RH discussion in Chapter 8.)

Taken in combination with temperature, the "comfort zone" is considered to be 20 to 60 percent RH with temperatures between 73 and 77° F. (23° to 25° C).

Sampling techniques for temperature and humidity. Measurements should be taken in a number of locations and times of day, particularly where workers complain that the area is too hot or too cold. Ambient temperature data from thermostats should not be accepted unless the investigator is satisfied that it has been cleaned and calibrated recently. Equipment preference is a desk thermometer and a relative humidity meter. Wet bulb, dry bulb thermometers can be used, but that degree of accuracy is generally not necessary.

Air Flow and Outside Air

As discussed in Chapter 10, ASHRAE 62-1989 offers guidelines for the amount of outdoor air recommended for a wide variety of institutional, commercial and industrial facilities. Table 2 from the 62 standard, a portion of which is shown in Chapter 10, recommends the cfm/occupant for various areas of a facility. 62 also offers an indoor air quality procedure for subjectively judging the outside air needs.

Higher ventilation rates are recommended for areas where smoking is permitted. Localized ventilation with direct outside exhaust may prove helpful in controlling contaminants at the source.

Inspectors should be aware that the amount of outside air may not be indicative of the level reaching occupants. The direction of air flow and it effectiveness in reaching the occupants are critical.

Measuring air flow. At this stage of the investigation, air flow is measured to be sure vents are functioning properly and the airflow is suitably directed. Exact measurements are not critical. Air movement from vents can be checked with smoke tubes, which can be obtained from safety supply houses.

Table 5-2 summarizes the temperature, humidity and air flow guidelines for occupant comfort.

TABLE 5-2. PERFORMANCE CRITERIA FOR MAINTAINING THERMAL ENVIRONMENT COMFORT CONDITIONS

PARAMETER	GUIDELINES
Operative temperature (summer)	73 to 79°F (at 50% RH)
Operative temperature (winter)	68.5 to 76°F (30% humidity)
Dew point	> 35°F(winter);< 62°F(summer)
Air movement	≤ 30 FPM (winter); ≤ 50 FPM (summer)
Vertical temperature gradient	Not > 5°F at 4" and 67"

Carbon Dioxide (CO_2)

Carbon dioxide (CO_2) is exhaled by the occupants in a room. Unless concentration reach exceptionally high levels, CO_2 is not considered a contaminant. It can, however, act as an excellent surrogate for measuring the concentrations of difficult to measure contaminants.

Outdoor ambient concentration of carbon dioxide are usually in the 250-350 parts per million (ppm) range. Experience has shown that occupants are apt to experience headaches, fatigue, eye and respiratory tract irritation if indoor levels exceed the outside level by 3-4 times, or 1000 ppm. This concentration itself is not responsible for the complaints. Rather, it is an indicator that other contaminants in the building have increased to undesirable levels.

CO_2 is used to define a lower limit of outdoor air needed to ventilate a building. 1000 ppm of CO_2 is the basis of the ASHRAE 62-1989 cfm requirements. CO_2 above 1000 ppm is not necessarily indicative that the building is hazardous or that it should be vacated. It just sends a signal that the ventilation is probably inadequate.

Sampling techniques for carbon dioxide. Detector tubes, which indicate CO_2 concentration by the length of color change on a sampling tube, affords inspectors an easy means of determining CO_2 levels. The detector tubes are available from industrial safety equipment supply houses.

Sampling locations should be planned ahead and should include areas where problems are suspected and areas where they are not. If no areas have been specifically identified as problem locations, then sampling locations should be selected from each area of the building. Information from a range of locations provides valuable data on the distribution of the air.

An error is often incurred by only sampling CO_2 at random times during the day. Baseline samples should be taken at all major locations first thing in the morning. An outdoor sample should be taken at the same time. Samples should be taken just before lunch, especially if many leave the building for lunch, and at the end of the day. The activities involved, the amount of foot traffic in and out, and the number of sampling locations will influence the additional number of samples to be taken.

CO_2 readings should be recorded by the specific location and the time of day. The pattern by locality can provide skilled investigators valuable information.

INTERPRETING RESULTS

Nothing more than common sense is required to interpret and act on some of the Phase 2 findings. If temperature, humidity or air flow readings compare unfavorably with the guidelines, then the system should be adjusted to bring the conditioned space into compliance.

CO_2 readings require more interpretation, but some crude guidelines can prove helpful. If indoor and outdoor CO_2 readings are about the same and remain that way throughout the day, chances are the amount of outside air entering the facility is adequate. The problem, instead, may come from an imbalance in the ventilation system or with the temperature/humidity.

If CO_2 levels are about the same as those outdoors in the morning but rise during the day, the air exchange rate for a 24 hour period is probably all right but not sufficient for the current level of occupancy or the activities taking place during the day. If the CO_2 level starts out unacceptably high, this is generally a sign of under ventilation. The system may have been shut down too early on the previous day (known in energy parlance as "coasting"), or the system may not have been turned on in the morning long enough before occupancy.

These interpretations of CO_2 vis a vis ventilation are grossly simplified using whole building averages. But knowing CO_2 for the whole building only tells part of the story. Distribution problems have been identified in up to two-thirds of the SBS buildings investigated; so it is important to look at ventilation in specific areas as well. Ventilation testing should not be limited to, and should never be reported solely as, whole building averages. Knowing the value for the whole building is useful information, but it can mask critical space-to-space variations. Ventilation rates within a building can vary by factors of five to ten. "Open space" offices and schools that have been partitioned are likely candidates for dead air pockets.

The importance of ventilation effectiveness, the ratio of outside air to the amount of air reaching the occupant zone, is discussed in the chapter on thermal comfort. To the occupant, it really doesn't matter how much air is entering the building if it is not reaching them. Testing ventilation rates in a variety of locations, where the people are, gets to the places where the complaints originate.

In using CO_2 to interpret contaminant levels, we also need to remind ourselves that CO_2 is essentially an occupant-based measurement and some contaminants are not occupant related. In a room with few people, for example, the CO_2 could be well below accepted levels while concentrations of some VOCs could be unacceptably high.

In other instances, CO_2 as a surrogate may indicate higher levels of contaminants, but increased ventilation may not be the best solution. In almost all cases, removal or containment at the source is preferable.

USING OUTSIDE PHASE 2 SUPPORT EFFECTIVELY

If an outside team is used in Phase 2, the diagnostic team will first seek to identify the nature of existing (or potential) IAQ problems, and within that context define the scope and objectives of their investigation. They will recap many of the activities described in Phase 1 and interview key personnel.

Management should make someone available to the team to expedite their work. It takes valuable time and money for the team to fumble through files or to try to find someone who knows about a piece of equipment. An individual who knows the organization and the facility should be available to the team to gather data, coordinate meetings and interviews, and support the team's investigation. Personnel to be made available for interviews/meetings should include senior administration and health/safety officers to review administrative matters, and the facilities manager plus technical and O&M staff to review facility-related concerns. Interaction with the occupants during the team's first inspection of the facility may prove valuable and management should encourage the team to engage in such dialogue.

This diagnostic team walk through is frequently a scoping audit to develope hypotheses regarding problem causes, how they appear to be manifested and relationships to building, equipment and O&M characteristics. Measurements already discussed; i.e., temperature, RH, air flow and possibly CO_2, will usually be taken.

From this information, the team will generally formulate a preliminary diagnostic procedure predicated on need, feasibility and cost-effectiveness.

The team usually concludes this phase with a meeting with management to review activities to date and present preliminary findings. Actions that can be taken without further diagnosis or analysis are recommended, such as O&M procedures, simple adjustments to the ventilation system, and/or changes in occupant behavior. If the team believes additional diagnosis will be needed, it may discuss the nature of that work at this time.

By the end of Phase 2, approximately 80 percent of the problems causing SBS or BRI have been identified. If problems still persist, then simple and complex diagnostic procedures will be needed. From here on, however, the going gets harder and the costs start to mount. And even with a capable team's exhaustive efforts, currently about 10 percent of the causes are never found.

PHASE 3. SIMPLE DIAGNOSIS

If this phase constitutes an outside team's introduction to the facility, it will most likely review the work done by the in-house staff and then proceed to do all, or almost all, of the work just described in Phase 2.

In addition more analytical procedures will be conducted, on site and back at the team's offices. The characterization of the symptoms, complaints and problems will usually influence other evaluation procedures. Medical evaluation may be involved. Typically, the major focus at this stage will usually be an engineering analysis of the HVAC system, the controls and the overall building performance to help establish environmental performance criteria.

Measurements may be taken of implicated pollutants. For example, if observable indicators (symptoms and building assessment data) point to carbon monoxide as a contaminant; then, instantaneous and periodic readings may be conducted using CO detectors and CO calorimetric indicator tubes with a hand pump.

Various diagnostic teams may use other measurements, usually as surrogates, during the simple diagnosis phase. Their selection is predicated on the type of facility and the information gleaned from the scoping audit. For example, certain contaminants, such as toluene, can be used in an office building as a surrogate for the intensity of human activity and the related adequacy of the ventilation system.

More precise data may be needed on the ventilation system. CO_2, as one means of measuring ventilation was discussed under Phase 2. Ventilation is also measured with tracer gas. CO_2 or tracer gas measurements can qualitatively and quantitatively verify the need to adjust or modify the ventilation system. Using either approach the system deficiencies are easy to evaluate and sometimes easy to remedy. Figure 5-2 offers a simple schematic of the single tracer gas technique for measuring infiltration and ventilation performance in a large building.

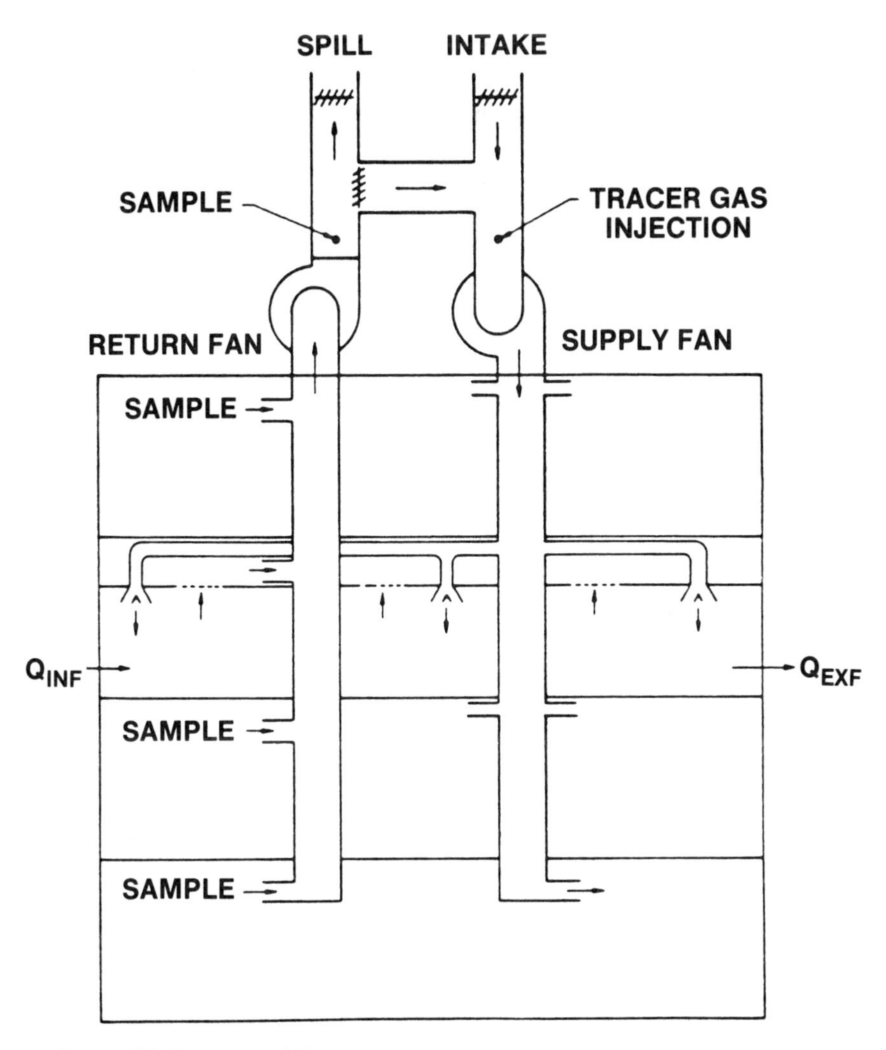

Source: U.S. Department of Energy

FIGURE 5-2. Single Tracer Gas Technique For Measuring infiltration And Ventilation Performance In Large Buildings

As previously noted, it is important to know what is going on throughout a building, and not just rely on some over-all air exchange value. In addition, as the diagnostics become more sophisticated correlating the simultaneous ventilation rate measurement to reported contaminant concentrations in a given area offers a much better picture of what is actually happening around the occupants in that location.

Researchers frequently use gas chromatograph (GC) and mass spectrometer (MS) tracer systems in conjunction. Chromatography, around since the 1950s, has become a good tool with the development of synthetic adsorbents suited to organic compounds and has proved useful when scientists have a rough idea of what they are looking for. A more sophisticated analysis device is the mass spectrometer, which can identify chemical compounds in the air by their atomic fragment "fingerprints." The United States Department of Energy cites the relative advantages of each system in Table 5-3.

TABLE 5-3. GC AND MS TRACER SYSTEMS

RELATIVE ADVANTAGES
[ ] GAS CHROMATOGRAPH (GS)
– High sensitivity - low tracer usage
– Lower unit cost - distributed systems
– Application: large buildings; steady-state measurements
[ ] MASS SPECTROMETER (MS)
– Mass range 1-300 amu - wide range of compounds
– Fast response time - active control
– Application: small buildings; dynamic measurements

Figure 5-3 depicts a multi-gas tracer measurement system using an MS system.

All in all, this "simple" diagnostic approach may require quite sophisticated simulation techniques to reflect the real life dynamic of building performance as well as tests for implicated pollutants or as surrogates. It requires a diagnostic team with broad expertise. From the management

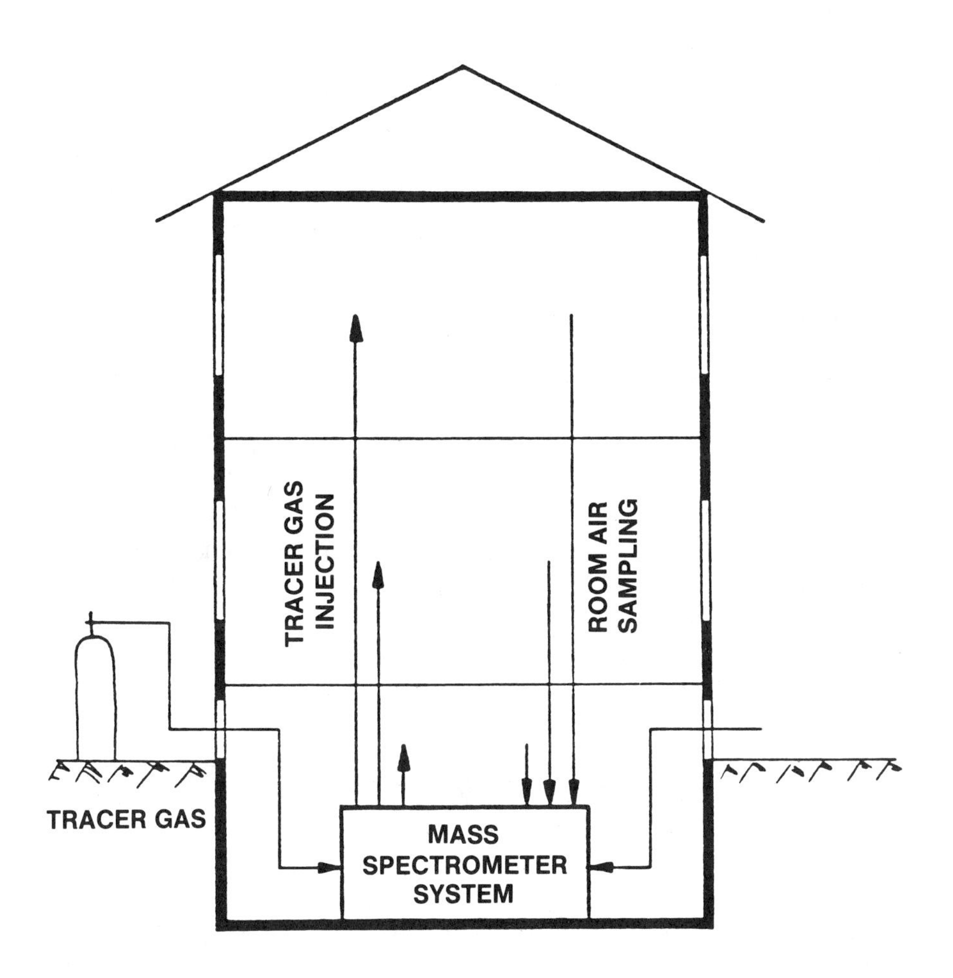

Source: U.S. Department of Energy

FIGURE 5-3. MULTI GAS TRACER MEASUREMENT SYSTEM

perspective, it is important that the team offer a multidisciplinary approach in its simple diagnostic procedures. This is discussed in greater detail later in the chapter under "Selecting a Diagnostic Team."

The results from Phase 3 often finds diagnostic team recommendations indicating that the original HVAC design was inadequate, or the system can no longer keep pace with changes in the facility or its use. In either case, if the design is inadequate to cope with the contaminant or thermal loads, the remedy is usually HVAC system modification. Recommendations for source control are likely. It is not unusual, even at this stage, for more operations and maintenance measures to be recommended.

PHASE 4. COMPLEX DIAGNOSTICS

If the implemented recommendations from Phase 3 do not resolve the problem, then more complex diagnostic steps are required. Or, Phase 4 measurements and analyses may be required to respond to litigation proceedings. Occasionally, they are needed to assess conformance to certain performance criteria.

This phase can be time consuming, costly and complicated. It should not be undertaken until other alternatives have been exhausted. During this phase, objective measurements of chemical, physical and microbiological parameters are paired with subjective responses from the occupants to their environment. Specific tests and real time measurements may be needed in a range of locations.

These locations usually include sites of complaints, areas with susceptible occupants, areas where no complaints have been registered, and an outdoor control site. Other areas that may be tested may include offices connected to the complaint area by the ventilation system, or areas with varying levels of ventilation efficiency.

Management should ask for and expect explanations in lay terms regarding analytical procedures to be used and procedures for selecting and applying appropriate sampling methods. The team should also be requested to describe methods it intends to use to provide quality assurance/quality control that will assure an acceptable level of accuracy, precision and reliability. The organization's spokesman and people responsible for internal communications should be fully apprised of actions to be taken and be able to explain them accurately in lay terms.

Recommendations from the complex diagnostics may call for a range of mitigating actions including, source control, system modifications, and maintenance and operational procedural changes.

PHASE 5. MONITORING AND RECURRENCE PREVENTION

Once an indoor air problem is resolved and the complaints are only a faint memory, the job is not over. One has only to recall the hysteria and harangues, not to mention the lost productivity and the investigation costs, to resolve not to get caught in such a bind again. An IAQ program, as described in Chapter 3, with an integral preventive maintenance program, as addressed in Chapter 7, helps to avoid such a dilemma.

In addition, monitoring procedures through routine checks and periodic measurements can help assure the program and the remedial measures are working.

"Monitoring" is sometimes used in IAQ literature in relation to investigative procedures, measurements and equipment use, such as radon monitors. When used in that sense, monitoring denotes activities prior to mitigation. Phase 5 monitoring is used in the broader, more conventional, sense of assuring informal compliance to some measure of acceptability following remedial action. Phase 5 monitoring is designed to check up on and keep track of measures taken to be sure beneficial results are preserved.

MEASURES OF ACCEPTABILITY

At this point, it helps to know just a little about acceptable levels, how pollutants are measured and the problems associated with measuring contaminants.

From the safe falling from a cliff toward the roadrunner in the cartoons to the fuselage explosion due to a faulty cargo hold latch to a particulate that migrates into the deepest reaches of an employee's lung, all can be lethal. All these scenarios involve increasingly smaller objects and increasingly more difficult sources to identify. We are left with no doubt where the ACME safe came from or when. In the case of an airborne contaminant, however, we may not know when exposure occurred, the length of that exposure or the contaminant concentration.

Furthermore, we know the cartoon character, who was squashed flat in the last frame, will be up devising new diabolical schemes in the next. In contrast, pinpointing the extent of damage from a particulate or even knowing if the damage was caused by an airborne contaminant is difficult, sometimes impossible.

The health effects may not become apparent until long after exposure. IAQ measurement is a subtle science made more difficult by many factors operating at once, including the exposure times, frequency and dose. To complicate things further, a particulate, for example, may have picked up a passenger or two, like radon daughters or VOCs. Intervening and confounding variables may make the identification of single contaminant very difficult and its source nearly impossible.

To make matters worse, there is a lot we still don't know about acceptable levels. And even if we were able to obtain reliable measurements, we aren't always sure what we've found out. Some data are available on acute exposure to certain pollutants in an industrial setting. Far less may be known about chronic exposure at lower levels in an office building. Our limited understanding of the damage some contaminants can do over a long period of time is the reason elimination at the source is much preferred to control by ventilation.

Other factors, such as age, general health or sensitivity, may play a greater role for some individuals than dosage or exposure duration. In many situations, children and the elderly are more vulnerable to a contaminant.

In many instances procedures for measuring a contaminant have been limited to research settings. Some techniques/equipment are very expensive. Few people are well-grounded in effective measurement procedures. Once data are obtained, only a limited number of analysts can accurately interpret those findings.

In this milieu of uncertainty, establishing measures of acceptability and knowing with confidence that they are being met is hardly an exact science. Nevertheless, acceptable levels can prove helpful. Some contaminants and accepted air quality guidelines/standards depicted in their usual units of measure are shown in Table 5-4.

TABLE 5-4. AIR QUALITY GUIDELINES/STANDARDS

Contaminant	Concentration	Avg. Time	Association Agency
Carbon Monoxide (CO)	10 mg/m^3=9ppm	8 hrs.	National Ambient Air Quality Standards (NAAQS)
Formaldehyde (HCHO)	120 µg/m^3 = 100ppb	continuous	American Society of Heating, Refrigerating and Air-Conditioning Engineers, Inc. (ASHRAE); American Industrial Hygiene Assoc. (AIHA)
Nitrogen Dioxide (NO_2)	100µg/m^3 = 50ppb	1 yr.	NAAQS
Radon	0.01 WL = 2pCi/L *	1 yr.	ASHRAE
	0.04 WL= 2pCi/L	1 yr.	National Council on Radiation Protection
	4 pCi/L	1 yr.	EPA
Total Suspended Particles (TSP)	75 µg/m3	1 yr.	NAAQS (Primary)
	60 µg/m3		(Secondary)

* Conversion of Working Levels (WL) to picoCuries/liter (pCi/L) assumes 50 percent equilibrium

ASHRAE 62-1989 offers information on various air quality standards and Appendix C of the 62 standard presents several tables depicting international, national and Canadian standards.

INVESTIGATION DIFFICULTIES

In order to have a better understanding of the investigative process and what is doable, it helps to have some appreciation of the difficulties inherent in the process. Some of the same concerns that impinge on setting acceptable levels also intrude on investigation procedures.

MULTIFACTORIAL CONCERNS

First of all, many SBS sites are not provoked by a single contaminant. Honeywell has found that 35 percent of the SBS buildings are also BRIs indicating bacteria, fungi and/or viruses are present along with other contaminants.

To complicate things further, pollutants may act independently, antagonistically or synergistically and little is known about these effects. Furthermore, other environmental factors, such as noise, lighting, temperature, humidity and ergonomics, can exacerbate contaminant problems. In addition, psychological factors, including job satisfaction and stress, are thought to affect some symptoms and certainly contribute to complaints.

THE HUMAN DIMENSION

For an indoor air problem to exist, there must be a receptor that is unfavorably affected by an airborne agent. Humans as receptors do not respond uniformly. Age, gender, nutrition, susceptibility, general health, ability to adapt and thresholds of sensitivity vary among individuals. All these factors may act, separately or in combination, as confounding variables. These variables may be further influenced by the comfort expectation held by the individual.

Not only do human have different expectations and varying reactions, they will report those concerns differently. Gleaning information from occupants is not without its problems. Investigations, for instance, are almost always prompted by what is perceived to be a problem by somebody. It will come as no surprise to students of human nature that one complaint, or the knowledge of some exposure to a potentially harmful substance (even at very low levels), will very likely foster additional complaints. A reporting bias will usually come from what Gable et al. call "epidemic psychogenic illness." Separating out the effects of interactions of low levels of exposure from "hysterical" behavior is exceedingly difficult.

KNOWING WHAT'S "RIGHT"

With regard to the contaminants themselves, little is known about acceptable levels of most pollutants. Even less appears to be known about low concentrations imposed over prolonged periods of time.

In many instances, the equipment, the measurement, the interpretation of the findings, or all three are far from perfect.

While the technology of measuring indoor pollutants has leaped ahead in the past decade, testing is still in its infancy for many contaminants. Even though we can now obtain detection levels of one part per quadrillion, testing equipment available on the market is frequently costly and/or may be unreliable, as with some radon testing devices. In addition, talented professionals qualified to administer and interpret results are in short supply.

From a more scientific perspective, any guidance for management is seriously limited by the dearth of studies about healthy buildings and the lack of any control buildings in most research studies.

In the final analysis, the difficulties associated with investigating indoor air quality problems are evidenced in the number that remain undetected. NIOSH reported in 1987 that 12 percent of their cases remained unresolved. Having investigated more than 2000 indoor complaints since 1976, the Ontario Ministries of Labour and Government Services noted in 1989, "In many cases, the causes of the symptoms could not be found."

Establishing any type of baseline or standard protocol has been thwarted by a lack of data, lack of uniformity in approach and the diversity of the disciplines involved. Each discipline views the problem/solution through eyes constrained by training and experience. To borrow an old line: The carpenter sees every problem as a nail and every solution as a hammer. It is this diversity that makes consultant selection so critical.

SELECTING A DIAGNOSTIC TEAM

A single consultant might be retained to develop a program, design an HVAC system, brief staff, train O&M personnel or perform a myriad of other IAQ related services that cannot be handled in-house by some organizations. But hiring one consultant to investigate an IAQ problem is almost always a serious mistake.

Unless management has narrowed its concern to one contaminant; e.g., radon, the multifactorial nature of sick buildings requires broader input. Experience has shown that standard epidemiology and industrial hygiene evaluation techniques may be inconclusive.

An epidemiologist, for example, usually sees the essential starting point in any building-related outbreak of illnesses as the complete characterization of the symptoms, complaints and their patterns. The industrial hygienist, following successful industrial research patterns, is inclined to test for specific pollutants. The engineer, who wants to get to the source of the problem, is apt to set aside the human research and head for the HVAC system.

Elements of every approach are important; and, while each has merit, the problem is frequently broader than the discipline. After costly testing by an industrial hygienist (IH), for instance, a contaminant in an office may be found to be well below accepted industrial exposure levels and the management is advised that the quality of the indoor air is acceptable. Unfortunately, we have found that we are playing by different rules and acceptable levels for contaminants in an industrial work place do not transfer to other work locations. Or, the problem may be that other contaminants are at work and this focused approach doesn't tell the whole story. In either case, after expending time and money, the management is given a clean bill of health by the IH, but the symptoms persist. Management says, on good authority, that a problem doesn't exist. Employees think management is playing games.

Every situation, every building is different, but selecting a diagnostic team is a fairly uniform process. The management needs to identify general task(s) that need to be done, set forth the criteria; e.g., experience, and ask for proposer's credential to meet those objectives. Experience in IAQ investigations is essential. Experience in similar facilities is desirable.

Except in general terms, it is very difficulty for the administration to describe exactly what is to be done. After all, if it was possible to be that precise, the problem would already be identified. It is imperative, therefore, that the references of prospective teams be requested and that they be checked out.

As with other contracts, it's also important to control assignability. The annals are full of stories of administrators, who thought they were getting experienced senior personnel, only to find the work being done without supervision by the June graduate. In a relatively new field, like IAQ, where talented professionals are still sparse, assignability becomes a critical consideration.

Potential diagnostic teams should be asked to describe in lay language what they propose to do and how they propose to do it; then, they should offer estimated costs for the proposed work. It is advisable to get scope of work and costs in advance of each phase. The earlier steps are more routine and easier to cost out. The more complex the investigation process becomes, the harder it is to anticipate all the procedures that will be needed and their associated costs. Fortunately, by the time the complex diagnostic stage is reached, management is familiar with and comfortable with the diagnostic team; so the uneasiness that comes with wading into uncharted waters can be avoided. If the comfort and trust isn't there, serious consideration should be given to getting another diagnostic team.

Applied evaluation of IAQ is an emerging field. Procedures traditionally used in industrial hygiene are inappropriate for non-industrial applications. Ignoring the wealth of information IHs have, however, would be foolish. Similarly, neither the medical profession, nor the engineers have the breadth of training and experience to go it alone. Ultimately, the team must provide a comprehensive and cost-effective approach to either resolving or preventing building-related problems. An effective team integrates the disciplines of science and engineering along with subtle interactions between occupants, buildings systems, and facilities management.

Even the best of teams, however, may never solve the problem. After spending weeks, even months, and countless thousands of dollars, they may come up empty handed. As depressing as all this sounds, it is comforting to remember that most sources are known and the means to prevent most problems are well established.

Chapter 6

Controlling Indoor Air Problems

In practice, there is little to distinguish controlling contaminants from their treatment. Treating an existing problem through mitigation or remediation is often the same exercise as preventive measures taken to control it in the first place. And once treated, the same actions become control procedures to prevent a recurrence. Some preventive actions, of course, can be taken that are not available in remedial treatment. Building design is an example.

Unless specifically mentioned, as in new construction, prevention and control strategies will be considered as one and the same in the following discussion.

TOXICITY AND HAZARD

Managing indoor air quality requires a basic understanding of toxicity and hazard. These terms are frequently used incorrectly in the press. They are not synonymous. Every substance is toxic to humans in some manner, or to some degree, even orange juice. A toxic substance becomes hazardous by intensity of the dose, the length of exposure or the manner in which it is introduced into the body. Few would fear orange juice in a glass, but no one would want it injected into his or her blood stream.

Hazard is actually a measure of risk or the probability that an unwanted event will occur. Hazard is usually measured in severity or magnitude. It is a product of toxicity and exposure (or dose). Exposure is usually defined as acute when it is intense for short duration, and chronic in the case of low level over a long period of time. Control procedures reduce or eliminate the level of toxicity, the dose or both; thereby, reducing the hazard, or risk.

MEASURING CONTAMINANTS

In the process of controlling a contaminant, if it can't be eliminated, then some gauge of acceptable levels is desirable. Building owners and facility managers, in most instances, simply do not need to know all the particulars about acceptable levels, measurement techniques, or the jargon that goes with it. When management gets embroiled in a serious problem, they soon learn more than they ever wanted to know about that particular pollutant. The Selected References and the references presented with each contaminant in Appendix A provide the resources to pursue such specifics.

Fortunately, about 80 percent of the problems associated with indoor air quality can be identified with a walk-through inspection using a smoke pencil, perhaps temperature and humidity gauges, and an educated eye. Extensive testing is used in approximately 20 percent of the SBS investigations and the cause is found in only about one-half of those. The difficulties associated with investigations, as discussed in the preceding chapter, help explain why roughly 10 percent are never found. Testing for contaminants clearly meets the test of diminishing returns: it is progressively more expensive while the possibility of results markedly declines.

The difficulties inherent in measuring contaminants, the limited talent available to analyze findings, and the high cost of measurement argue for an effective control program. Since such a large percentage of the problems can be found in a relatively simple walk through, control opportunities are relatively easy to identify. Many only take a little common sense. If investigations and interviews, for example, reveal three findings: (1) occupants evidence the symptoms associated with bioaerosols; (2) the area was recently flooded; and (3) the carpet got wet, the logical answer is: replace the carpet. The alternative, in almost every case, would be to go through an exceedingly costly process of testing for bioaerosols, only to end up with a recommendation: replace the carpet.

CONTROLS

To accommodate occupant needs, building owners and facility managers must recognize the multiple purposes of controlling the quality of indoor air. First of all, federal requirements for specific contaminants must be met.

Furthermore, satisfying guidelines that do not have the force of law may still be legally prudent as a defense against potential lawsuits. The underlying purpose in all control procedures is to be sure the indoor air:

- maintains the quality needed for safety and health;
- satisfies comfort and productivity needs; and
- is as cost-effective as possible.

The variability of organizational and occupant needs further complicate control procedures. The "status quo" is never static. As facility managers and operating engineers have long recognized, seasonal variations call for accommodations in air conditioning (heating and cooling) and in lighting. In fact, controls to meet these needs vary from day to day, even hour to hour. Similarly, air quality needs may vary by tasks and the scheduling of those tasks. For example, VDT use may be much greater at the end of the month. Or, copiers may suddenly be used extensively just before the big meeting.

Obviously, operators could be run ragged trying to meet every little variation. The goal is to avoid rigidity in implementing control strategies and to allow sufficient flexibility to meet these variables. As more sophisticated automated systems emerge, their sensing/control devices will facilitate responses to individual needs and conditions.

CONTROL METHODS

Methods for controlling contaminants fall loosely into four categories; (1) elimination at the source(s), (2) ventilation, (3) design, and (4) operations and maintenance procedures. They can be further broken into more precise measures to help identify the needed control actions. Many control procedures were first developed in the industrial workplace. While industrial hygiene and safety considerations are beyond the purview of this book, some are mentioned here because of their carry-over implications to other occupational settings.

1. Elimination - the complete removal of; (a) the biological agent, (b) a toxic substance, (c) a hazardous condition and/or, (d) the source. Elimination procedures include maintenance actions to remove the breeding grounds for bioaerosols, the removal of friable asbestos, or the banning of smoking.

2. Substitution - the deliberate purchase or use of less hazardous materials; e.g., pesticide selection, the purchase of furnishings or building materials, the selection of latex- based paints over oil-based paints wherever possible.

3. Isolation - containment, encapsulation, shielding, sealing, and the use of distance are all means of isolating a contaminant or a source from exposure to humans. Distancing may be accomplished through location and/or time of use. Examples of isolation controls include painting during unoccupied hours, asbestos encapsulation, and removing ornamental plants from the facility before spraying with pesticides.

4. New construction/renovation design - many design steps can be taken to prevent problems from occurring including ventilation effectiveness, thermal comfort, lighting, the selection of building materials and maintenance needs. Filter access, for example, is an often overlooked but critical design consideration.

   Procedures to commission a new building, including "bake-outs," air purging and occupancy schedules, should be considered.

   Modifications to buildings to be avoided include changes that restrict original air flow design, such as partitions in an open space facility, or changes that increase contaminant load beyond HVAC capabilities.

5. Product or process change - while borrowed from industry, change in process controls have applications in other settings. Component design to reduce emissions might be used, for example, to affect the way a Zamboni machine is exhausted at an ice rink.

6. Housekeeping and dust suppression - actions that keep surfaces clean of contaminants, prevent their redispersion, and/or eliminate personal contact entirely are important control measures. Some very common place controls are windbreaks, care in preventing vacuum cleaner leaks, and efforts to contain dusty sources, such as coal bins.

7. Maintenance and work practices - specifications for the proper work procedures to reduce or control contaminant releases, such as

pesticides, need to be spelled out and should be part of training procedures. Maintenance practices (many of which cut across other control procedures listed here) are vital, especially in the automated control and HVAC areas.

8. Replacement - insulation, carpeting, wall coverings, etc., which when wet can serve as breeding grounds for microorganisms, need to be checked regularly and replaced when damaged.

9. Education, training, labeling and warning procedures - some training, labeling and warning procedures are required by law. Whether required or not, workers and management must be educated as to the nature of hazardous materials and ways to minimize risk in their use. In some instances, educating building occupants, guests or public may be necessary.

   Owners may need to force some education on professionals, through specification or professional qualifications. The owner carries the ultimate responsibility to see that architects and engineers use appropriate designs and select safe building materials.

10. Sanitary procedures and personal protective devices - the use of hygienic principles to reduce or eliminate hazardous materials from a person may be critical under certain conditions. In a hospital setting protective procedures are accomplished through clothing changes, showering, chlorination, etc. Protective devices are usually used where other control devices are not technically or economically feasible; e.g., respiratory protective devices when O&M people are working with fiberglass.

11. Storage and disposal - laws prescribe the storage and disposal of some contaminants. Care should always be taken to use and store materials as indicated on the label. When in the slightest doubt as to proper use, storage or disposal, good control practices dictate contacting manufacturers, state and/or federal agencies first.

12. Filtering and purification - the use of filters and purification devices in the air distribution system, with outdoor air and mechanically recirculated air, is an essential control factor. Filters & purification

devices appropriate to the need should be used and maintained/ replaced on a regularly scheduled basis.

13. Ventilation - through increased outside air or exhaust with make-up ai Dilution is a preferred control when the contaminant/source is unknown, source treatment is too costly, as an intermediate step during investigation, or when the source is localized. Ventilation control is more than the amount of outside air brought into a facility, and includes the quality of outside air, the effectiveness with which it reaches occupants, and the efficiency with which it reduces contaminant levels. The opportunities and limitations of ventilation as a control are treated in greater depth in Chapter 8.

Management may also wish to seek the advice and counsel of medical experts as a control support through medical surveillance or treatment. Medical surveillance may include occupant preplacement screening that will restrict high risk persons or provide medical exclusion; e.g., reassignment by location or task of chemically sensitive individuals. Building related illnesses (BRI) by definition depend on medical verification, which in turn usually suggests the cause and the control measures to be taken.

Overriding all of these control procedures are some administrative and management considerations as well as judgment calls. Personnel rotation or scheduled relief breaks may be necessary to reduce the time of exposure to an essential process. As an example, equipment requiring relatively high power charges, such as computers, put positive and negative charges on dust particles; thus, attracting some while pushing others in the face of the operator. That's why computer and TV screens get dusty even on the lower side of convex surfaces. Unfortunately, the same amount of dust is forced into the air computer operators must breathe. These concerns and others related to the use of VDTs may require task rotation and consideration of female workers who are pregnant. To assure control procedures are planned and executed properly, formal organizational steps to delegate authority and responsibility must be made. These management considerations are discussed in Chapter 3.

SPECIAL CONTROL CONSIDERATIONS

The importance of thermal conditions to occupant comfort and the relationship to particular contaminants is well documented. The role of the HVAC system in delivering the required thermal conditions and meeting IAQ ventilation needs is a critical IAQ concern. The fact that the HVAC system is also a source of contaminants is a key consideration. Because of their unique characteristics and importance to IAQ, thermal comfort factors and ventilation are discussed separately in Chapter 8 and HVAC systems in Chapter 9.

Other control factors of particular concern are filter selection and maintenance, purchasing, and new construction/modifications design. These areas are treated very briefly below from the management perspective. Filters are also discussed in Chapter 9 as an aspect of HVAC systems.

FILTER SELECTION AND MAINTENANCE

The primary purpose of filters is to reduce the contaminants in the airflow. The effectiveness with which they fulfill this purpose is determined by the type of air cleaner used, location in the system and the amount of air which passes through. The more effective the air-cleaning system is in removing contaminants from the already conditioned return air, the more that air can be substituted for outdoor air; thus, reducing the energy costs. The amount of filterable material that can be removed is determined by the volume of flow through the cleaner and its effectiveness.

The selection of an air-cleaning system is based on the contaminants to be removed, such as the size of particles, and any specialized needs that must be met. For example, a special air washer system is used to remove carbon dioxide (CO_2) in the closed system of an atomic submarine; however, its high cost and maintenance needs make it impractical for commercial buildings. In other circumstances, the centrifugal separator is dependent on the density of the particles, their size and the density of the air. Since very small particles cause an aerodynamic drag in a moving airstream due to their surface area versus mass, centrifugal separators only have application in settings where larger particle removal is required, such as dust removal in industrial plants.

Building HVAC systems typically rely on media filters and electrostatic air cleaners. The media filter is a porous material that strains out particles as air is forced through it. Media filters range from the more crude "dust-stop" filters used in most warm-air furnaces to the high efficiency particulate air filter (HEPA). The HEPA filter is capable of removing submicron particles.

The effectiveness of media filters change as particulate matter builds up. If left alone, they gradually increase in collection efficiency. That, however, is not a good reason to leave them unattended. As they clog, the flow rate decreases. Resistance from clogged filters reduces air flow and system efficiency. Eventually, filters must be changed. The dirtier they are when changed, the worse the contamination is apt to be to the air flow during the changing. Media filters require regular maintenance with scheduled cleaning or replacement. Several years ago, the Galveston Independent School District initiated a monthly filter replacement program and found it to be cost-effective. System efficiency and better equipment operation resulting from the monthly filter changes actually realized the district a positive cash flow as the equipment and utility bill savings exceeded filter replacement costs.

Electrostatic air cleaners (EAC) have demonstrated very high efficiency in removing particles in the range 0.01 micron to 5.0 microns. The EAC acts by charging the particles passing through, causing them to be attracted to oppositely charged plates. The collected material must be regularly removed from the collector plates. EACs operate properly only when absolutely clean. Build up on plates not only reduces effectiveness, but the units can become microbial sources. In small installations, the whole filter is removed for washing. Large EACs are designed with wash-in-place capability; however, unless maintained properly, they can be a source of serious microbial contamination.

Gases and vapors can be filtered using adsorbers or absorbers. Activated carbon, with its large surface area due to voids within its structure, is a good example of this type of system. Adhesion of the gas molecules to the carbon removes the contaminants from the air. This process works fairly well on large molecules, such as odor molecules, but does not do an effective job of removing small molecules, such as carbon monoxide.

The source of contamination can influence the placement of filters in the system. If the source of contaminants is the conditioned space, the air cleaner can be placed in the recirculating airstream or the mixed airstream. If the main source is outside, then outdoor air filtration is a primary consideration.

The above discussion is presented to acquaint managers with the air-cleaning options as control measures and their maintenance needs. For more precise information, such as design considerations, the reader should refer to ASHRAE 52-1968 (RA 76). In general, ASHRAE efficiency ratings for filter types serve as good measures of efficiency for most bioaerosols.

Unless the staff has particular expertise in this area, it is wise to consult a specialist when filter design or modification decisions are needed.

PURCHASING

Caulks, sealants and adhesives have various levels of VOC emissions depending on the compound. Building, pipe and duct insulation with urea-formaldehyde insulation or asbestos-containing acoustical, fireproof and thermal insulation can have an adverse effect on IAQ. Paints have highly variable mixtures of VOCs and have high release rates during the short-term curing process. Many of the problems associated with these products can be avoided through selective purchasing.

The careful selection of construction materials and furnishings can help alleviate indoor air problems at the source; e.g., the level of formaldehyde emission from pressed wood products. Table 6-1 lists materials of particular concern that warrant careful selection procedures. Materials of concern encompass site preparation, envelope construction, mechanical system and interior finishing. The March 1989 and February 1990 issues of Indoor Air Quality Update have excellent sections on selecting and specifying IAQ sensitive building materials.

Reference in the specifications to certain federal or industry guidelines can avoid major headaches, both figuratively and literally, later on. In purchasing materials directly or through the contractor, standards can be specified so that materials meet federal regulations, such as CFR 24, Part 3280, or industry association guidelines (NPA 8-86 and HPMA FE-86). Manufacturer's Safety Data Sheets should be required of all vendors as per

OSHA mandate. An additional precaution is to ask that each chemical or material in a composition product be identified.

TABLE 6-1. BUILDING MATERIALS OF PARTICULAR CONCERN

SITE PREPARATION AND FOUNDATIONS

- soil treatment insecticides
- foundation waterproofing, especially oil derivatives

ENVELOPE

- wood preservatives
- concrete sealers
- curing agents
- caulking, sealants, glazing compounds and joint fillers
- insulation, thermal and acoustical
- fire proofing materials

MECHANICAL SYSTEMS

- duct sealants
- insulation of ducts

INTERIORS AND FINISHES

- subfloor or underlayment
- floor or carpet adhesives
- carpet backing or pad
- carpet or resilient flooring
- wall coverings
- adhesives
- paints, stains
- paneling
- partitions
- furnishings
- ceiling tiles

Adapted from Indoor Air Quality Update, Arlington, MA

NEW CONSTRUCTION/RENOVATION DESIGN

All materials pollute, some a little, some a lot, but they all contribute to a deterioration of the air quality. "Engineering out" materials and conditions known to have a particularly adverse effect on indoor air quality during new construction or renovation may prove to be the easiest control measure of all.

Now that it has become evident that a building can be hazardous to one's health, the reduction of the potential hazards become a critical criteria of building design and construction.

Put It in the Specs

The owner, who is ultimately responsible for the facility, cannot assume an architect or an engineer fully understands the design and materials implications associated with indoor air quality. It is incumbent upon the owner to include sufficient direction in the specifications to obtain a facility that is energy efficient, aesthetically pleasing, productive and hygienically safe. Appropriate IAQ specifications offer a quality control opportunity, that can save millions in remedial measures, and could lessen legal liability.

Facility Siting, Foundations

Siting of a facility should consider potentially negative influences on IAQ; e.g., the quality of the soil, particularly in radon affected areas. The proximity of roadways and vegetation may, depending on the type and placement, filter out contaminants or serve as a source of allergens.

The nature and location of polluting activities should affect design. For example, exhaust near a localized activity may serve as an effective control mechanism. If theatre or craft production, car repair work or maintenance operations are going to involve toxic gases or particulates from brazing, welding, cutting or soldering, then direct venting to the outside should be part of the design.

Precautions should be taken in new construction to thwart radon entry. Recommendations made by Brennan and Turner several years ago in "Defining Radon" still provide excellent control guidance. They recommended: (1) slabs and basements be poured with as few joints as possible and poured right to the wall; (2) wire reinforcement be used in slabs and walls to help prevent future cracks; (3) seams and perimeters be caulked with polyurethane; (4) dampproofing, sealing and coating the walls be done to slow entry; and (5) sub-slab drains and several inches of #2 stone be placed below the slab during construction to provide sub-slab ventilation potential.

Thermal Comfort Design

In addition to the consideration given to the thermal environment and humidity, air flow patterns need careful HVAC designer attention. Designers historically were concerned almost exclusively about temperature control. For years, humidity controls were ignored except in areas of extreme climate or special conditions. While we've provided "climate controlled" environments for rare books, masterpieces, and special documents for many years, we have only recently begun to recognize the importance of humidity in a comfortable, hygienically safe environment. The role humidity plays in fostering/controlling contaminants is not yet fully appreciated.

We have not progressed very far in making sure that the outside air or the cleaned, recirculated air is actually reaching the occupants. With the advent of sealed windows, occupants became totally dependent on the designer's understanding of air flow. To illustrate one renovation problem, we have only to look at the partitioned "open space" offices and schools of the 1970s where pockets of dead air have been created.

Ventilation Effectiveness Design

A more pervasive problem has been revealed in studies as recently as the mid-1980s, which found that most office buildings had both the supply air outlets and return air inlets located at ceiling level. This placement can cause a short circuiting of air. As the air pours across the ceiling, half of the room -- the half where the occupants are -- is left with poor ventilation.

An owner does not have to understand complex formulas on mixing efficiency to be able to spot grill work at the ceiling level and to see why the room still seems "stuffy" even with increased outside air. Such designs also severely constrain the value of ventilation as a control measure. Ventilation effectiveness, the ratio of outside air reaching occupants compared to total outside air supplied to the space, determines the capability of the supply air to limit the concentration of the contaminants. Design considerations are implicit in the discussions of temperature, humidity and ventilation effectiveness found in Chapter 8.

Maintenance Considerations

The importance of providing access for maintaining all equipment is discussed in Chapter 7. It cannot be overemphasized. Failure to allow for maintenance access occurs entirely too often. The resulting horror stories are legion. For owners and managers, this should be a critical design concern.

Sources of contaminants and their controls as discussed in this chapter and throughout the book suggest further design considerations. For example, the material on bioaerosols discusses the problems associated with duct linings. Burge has observed, "The general opinion in the bioaerosol field is that the millions of miles of fiberglass duct lining already in place constitutes a microbiological time bomb." Unless one feels comfortable sitting on time bombs, some consideration should be given to quality maintenance procedures (Chapter 7) or using an alternative to fiberglass duct lining in renovation and new construction design.

Bake Outs

Bake-outs are frequently suggested as part of the commissioning process to control VOC contaminants in new construction. Elevated temperatures speed VOC emissions. The bake-out is, therefore, expected to accelerate the off-gassing process prior to occupancy and thus reduce the pollution levels after the building is occupied. A bake-out, is controlled by adjusting three parameters; the duration of the bake-out, the indoor air temperature during the bake-out, and the ventilation rate during the process. In 1989, J. Girman and his colleagues (State of California) reported reduced VOCs as a result of the bake-out process. Researchers, headed by C. Bayer, at the Georgia Institute of Technology, reported in 1990 that total counts of VOCs were "about the same before and after the bake-out."

Emerging Technology

Finally, owners should keep their eye on emerging technology, particularly in the area of automated controls and ergonomically correct furniture. A leading manufacturer now has on the market control panels that allow each individual to regulate his or her own temperature, air motion and lighting. The unit also offers electrostatic air filters to assure a positive supply of cleaned air to each work station and the opportunity to add "white noise" to the environment so a barely discernable constant sound blocks out distracting noises. Weighed against personnel costs related to absenteeism and lost productivity, the physical and mental health offered by this self-controlled personal environment may be cost-effective. The manufacturer claims a 5 percent increase in productivity will yield a payback of less than two years; a 1 percent productivity increase offers a four to seven year payback (varying with personnel salaries).

FIGURE 6-1. PERSONAL ENVIRONMENT CONTROL EQUIPMENT

Managers can't keep up with the state-of-the-art on all technological fronts. Keeping tabs on the continuing emergence of controls, equipment and software to satisfy IAQ needs, reduce energy usage, and optimize productivity and space utilization would be almost impossible. The obvious benefits, however, warrant a call or two to control manufacturers as new construction/renovation planning begins.

CONTROL EVALUATION AND MONITORING

Control management requires some evaluation of the steps taken. The level of evaluation will vary with the condition(s) which caused the problem. With sick building syndrome (SBS), where the cause may never be identified, the cessation of complaints may serve as sufficient evaluation of treatment procedures. In the case of something as alarming as Legionnaire's disease, medical verification and an analysis of water sources for Legionella may be warranted.

Following the treatment for SBS or BRI, a critical component of effective control management is monitoring. Monitoring may range from a simple oversight procedure that assures continued adherence to control practices, to intermittent or continuous atmospheric sampling and analysis methods.

Sampling and analysis for hazards may include area, personal, process, or duct sampling to determine characteristics of emission, level of exposure and operational conditions. Sampling and analysis procedures are usually very specialized, conducted by outside consultants and are expensive. It is an evaluation procedure generally reserved for extreme circumstance or research.

Currently, the most accepted monitoring approach to assure continued quality of the indoor air is monitoring carbon dioxide (CO_2) measurements. As discussed in Chapters 4 and 5, CO_2 is frequently used as a surrogate for other contaminants. It is particularly useful in measuring concentrations of combustion products, including nitrogen dioxide (NO_2), carbon monoxide (CO), sulphur dioxide (SO_2) and sulphates (SO_4). CO_2 monitoring is valuable in assessing the effects loading docks, parking garages, service garages and train stations might have.

Except in instances where sources, such as vehicle exhaust, affect CO_2 content, the concentration of CO_2 in the outside air is fairly constant. With information about ambient CO_2 levels outdoors and what is going on indoors, we can learn a lot about the effectiveness of the ventilation system. The cfm/occupant requirements in ASHRAE 62-1989 are based on desirable CO_2 levels of 1000 ppm or less. Generation rates of CO_2 by people are related to metabolism and activity or less. If the CO_2 occupant generation rates and the ambient CO_2 level of outside air is known, then indoor CO_2 concentrations will indicate the air exchange rate; and CO_2 can serve as an indicator of the concentrations of most indoor contaminants.

If CO_2 sensors are used to monitor concentrations, care should be taken to be sure they are sufficiently reliable. If they drift more than 100 ppm per year, then they will not offer a reliable reading.

CONTROL BY CONTAMINANT

The effectiveness of specific controls varies by contaminant. Ventilation as a control device, for instance, will not benefit and may aggravate certain contaminant situations. When the pollutant is known, controls should be contaminant-specific. Some control measures are required by law. In such instances, legal reference is made to the applicable law or code in the following discussion of controls by specific contaminants.

ASBESTOS

Abatement methods are:

1) Operations and maintenance
2) Repair
3) Enclosure
4) Encapsulation
5) Removal

The Asbestos Hazard Emergency Response Act (AHERA), which applied to the schools when passed in 1986, may be extended to other facilities. The federal law specifies acceptable abatement procedures. Some people falsely assume that AHERA requires asbestos removal. The law only requires schools to exercise every effort to protect human health and the environment by "the least burdensome method." Financially, the least burdensome method seldom equates with removal. Furthermore, as the Harvard report, Summary of Symposium on Health Aspects of Exposure to Asbestos in Buildings, reminds us, "Removal itself is not without risk."

The threat of lawsuits has prompted building owners to opt for removal; however, removal has the potential to increase, rather than decrease, indoor air concentrations. Removal and disposal exposes some workers to high concentrations of airborne asbestos. Owners may inherit an even greater liability through removal, for they are responsible for the asbestos at the disposal site for 40 years.

Abatement specialists, fully trained with appropriate credentials and experience, should be employed for asbestos detection, encapsulation, removal, disposal and air monitoring procedures. Owners should seek protection by having specialists bonded and insured. State counterparts of EPA and OSHA can help identify appropriately qualified contractors. (See Chapter 10, States, for information on obtaining a state directory of IAQ contacts.)

When reviewing contractor credentials, try to retain the services of specialists with experience in like facilities. Contractors, who gained their experience in schools, may not appreciate the difficulties to be found in government offices, hospitals, or multi-tenant high rise buildings where the HVAC system can't be shut down or the space evacuated during work.

Asbestos removal, when the HVAC is not shut off, presents a whole new set of conditions. Asbestos removal generates high levels of asbestos dust that may be transported in the HVAC system. Preplanning, by a team which includes a mechanical/HVAC engineer with asbestos expertise and an industrial hygienist, will help determine the precautionary steps to take; e.g., work area isolation, negative air filtration, personnel and area decontamination. It may be necessary to weigh the viability of the continued operation of the HVAC system.

OSHA guidelines for asbestos removal can be found in Title 29, Code of Federal Regulations, 1910.1001 and 1926.58, U.S. Department of Labor, Occupational Safety and Health Administration, 1985.

BIOAEROSOLS

A review of microorganism and their sources suggests ways to control bioaerosols and microbial contaminants. Sources for biological growth include wet insulation, carpet, ceiling tile, wall coverings, furniture and stagnant water in air conditioners, dehumidifiers, humidifiers, cooling towers, drip pans and cooling coils of air handling units. People, pets, plants, insects, soil may carry biological agents into a facility or serve as potential sources. Frequently, bioaerosols settle in the ventilation system itself where viable spores and bacteria can incubate and grow. The inevitable dust and darkness inside duct work plus condensate moisture (often resulting from variations in temperature and humidity) can work together to turn the vast surface areas provided by fiberglass lining into breeding grounds for mold.

Most remedies fall in the area of maintenance and preventive maintenance; e.g., cleaning filters and wet areas in the ventilation system; replacing water-damaged carpets, insulation; maintaining relative humidity between 30-60 percent; cleaning/disinfecting drain pans and coils; etc.

While good quality filters will remove microbial contaminants, they will not do the job if the contaminants are generated downstream. Stories are told of "rattling the cage" and increasing contaminant levels in occupied space four-fold over outside air levels. Cleaning humidifiers, coil surfaces and drain pans are the most effective remedial actions for microbial contaminants generated downstream of the filter banks. Chapter 7 discusses bioaerosol-related maintenance.

Filtration of bioaerosols is relatively easy as microorganisms fall in a particle size range that can be effectively trapped by a variety of filters. Most microbial cells range from 1-20 μ with a few from 0.5-200 μ, all of which can be removed in a good quality filtration system. 50 percent atmospheric dust spot efficiency filters will remove most microbial particulates in the return or outdoor airstreams.

ASHRAE 62-1989 has added a discussion of biological contaminants to its earlier 1981 guidelines.

Biocides are sometimes recommended for cleaning ventilation systems. The American Conference of Governmental Industrial Hygienists

Bioaerosol Committee recommends that no biocide be used in an operating ventilation system. Decommissioned HVAC systems can be cleaned with biocides provided all biocide is removed before the system is restarted. Risks associated with biocides may be worse than exposure to microorganisms.

The need to replace duct lining is seldom required. If it is, however, it is often cheaper and easier to replace the duct section than just the lining.

COMBUSTION PRODUCTS

The best control of internally generated combustion products are efforts to maintain, properly adjust and carefully operate all combustion equipment. Vehicular exhaust from garages, loading docks, etc., constitute another major source of combustion products, that needs to be carefully managed.

The National Aeronautics and Space Administration (NASA) has found that house plants can serve as living air cleaners for the volatile organic chemicals often found in combustion products. Flowering plants like the gerbera daisy and chrysanthemums have been found to be particularly effective in removing benzene; the philodendron, spider plant and golden pothos in removing formaldehyde. Other plants found to be effective air purifiers included English ivy, ficus, Chinese evergreen, bamboo palm, peace lily, mass cane and mother-in-law's tongue. NASA also found that the root-soil area as well as the plant leaves act as purifiers.

When unusually high levels of combustion contaminants are expected in an area, additional ventilation can and should be used as a temporary measure.

Studies have shown that NO_2 can be removed from indoor air by processes other than air exchange. Materials can actually serve as significant sinks for NO_2. The judicious selection of construction materials and furnishings can effect NO_2 levels; e.g., some studies have shown that cement blocks can remove all NO_2 within 30 minutes. Humidity can influence the rate and mechanism of NO_2 removal for some materials.

ENVIRONMENTAL TOBACCO SMOKE

The control and treatment of ETS generally falls into five areas:

(1) remove the source - eliminate smoking. Decrees to eliminate smoking are policy decisions with employee relation considerations.

(2) modify the source - relocate/separate smokers. Separating smokers will reduce, but not eliminate, ETS. Chapters 3 or 7 discuss management and technical aspects of this control option.

(3) dilution ventilation. Increasing ventilation may prove to be the most desirable option, but it is a very expensive one. Ventilation rates to satisfy ETS-related health concerns have not been established. ASHRAE 62-1989 guidelines are designed to reduce tobacco smoke odor; not necessarily reduce health risks. Janssen, chairman of the ASHRAE committee for 62-1989 warns, "The great problem is that ventilation alone cannot effectively control the risk from ETS. Convection currents in a room may play a greater role than dilution with outdoor air."

(4) filter the contaminants. HEPA filters and electrostatic precipitators can remove respirable particles. Since both odor and irritation are mainly caused by the gaseous phase of smoke, this is not a particularly viable option. Granulated filter media, such as activated carbon, or some type of catalytic system can collect the volatile components of ETS. EPA has cautioned that the use of filters to remove ETS is technically and economically impractical. House plants have also been shown to be helpful in removing contaminants. (See sections on "Combustion Contaminants" and "Formaldehyde.")

(5) isolate smokers and exhaust contaminants directly to outside. Establish a smoking lounge, where the return air duct work is blocked and exhaust registers move air directly outside.

FORMALDEHYDE (HCHO)

Through selective purchasing, materials can be obtained with lower potential formaldehyde off-gassing. Researchers have found up to a 23-fold difference in emission from the same products from different manufacturers due to different resins being used and/or pre-treatment to reduce emission levels. Contact and inquiries with dealers and manufacturers with regard to potential formaldehyde emissions prior to purchase is warranted and may result in materials with lower exposure levels.

The potential to significantly reduce the irritation from formaldehyde warrants a pro-active, preventive approach. Laminated products should have all exposed surfaces or unused "plug holes" covered with laminate or replugged.

Pretreating carpet by heating or steaming can accelerate off-gassing. Similarly, "bake-out" procedures may accelerate off-gassing from other construction materials and furnishings. Purging with outside air, with the return air blocked, can air out the facility before occupancy. The effectiveness of the "bake-out" control procedure is currently under study.

Increased ventilation rates, within the typical indoor parameters, do not necessarily show any significant reduction in formaldehyde concentrations.

Barrier coatings and sealants might be used to reduce formaldehyde emissions. Researchers at Ball State University (Godish) have tested various sealants on particleboard flooring and have found they are effective in the short run and after six months. Barriers, such as vinyl floor coverings, have reduced HCHO in residences up to 60 percent even with other HCHO emissions present. Barrier coatings and sealants pose their own IAQ problems and adequate ventilation should be maintained during application and until the strong odor fades. Prior to using sealants, notification to chemically sensitized people is recommended.

A study by the National Aeronautics and Space Administration (NASA) has found ordinary house plants can significantly reduce levels of formaldehyde. NASA found philodendron, spider plant and golden pothos were most effective. The agency also determined that the root-soil zone was a vital part of removing contaminants and recommended maximizing air exposure to the plant root-soil area. For large volume areas, activated carbon filters containing fans should be an integrated part of any plan using houseplants according to NASA.

RADON

The primary sources of radon in buildings with high concentrations is the pressure driven flow of radon soil gas. This pressure difference may be the result of indoor-outdoor temperature differences, prevailing winds, the mechanical ventilation systems and combustion devices that have a depressurizing effect on the building.

The most effective means of controlling radon is to prevent it from entering the building. In radon areas, precautions should be taken to thwart radon entry. In existing buildings, sealing all cracks and openings around drains is helpful. Some studies have reported mixed success with this control procedure. Facilities with concrete block walls offer many, many avenues for radon entry. Sealing the block walls and ventilating the cavities have cut radon levels up to 90 percent.

In residences, mechanical ventilation of crawl spaces has proved to be very effective.

Using fans to draw soil gas from beneath a slab in sub-slab ventilation has been shown to be 50 to 90 percent effective in reducing radon levels.

In situations where a basement is fairly tight, pressurizing the basement can be effective. This relatively simple and inexpensive process has reduced radon concentrations 65-95 percent below radon concentration guidelines. As an alternative, increasing ventilation at the first floor level with exhaust out of the basement, as conducted by Wellford in his Pennsylvania study, proved to be effective.

General increased ventilation is frequently recommended in the literature as a way to reduce indoor radon concentrations. Repeated studies have found no correlation between radon concentrations and air exchange rates. (See Figure 1-1.) In other words, just increasing air exchange rates in general may not work. Ventilation must be applied in a specific fashion to be effective as a control.

Several states require that those who engage in, or profess to engage in, testing for radon gas or the abatement of radon gas be certified or licensed. The state public health, radiation protection office or environmental agency can provide information as to whether such credentials are required; and, if so, provide lists of those who meet the requirements. The EPA's Radon Measurement Proficiency Report, which lists firms and laboratories that have demonstrated their ability to accurately measure radon in homes, may be available from the state. The state agency can usually provide guidance regarding the possibility of radon contamination and effective treatment in a particular geographical area. The state can also offer counsel on the availability of effective detection devices and/or services.

RESPIRABLE PARTICULATES

Particles are the easiest contaminants to remove from the air stream. Media filters and electrostatic air cleaners are used in building HVAC systems. While more expensive to install and operate, high efficiency particulate air (HEPA) filters and electrostatic precipitators can remove respirable particulates from the air stream very efficiently.

Appropriate maintenance with scheduled cleaning and replacement of filters is essential. (See the discussion on filters earlier in this chapter and in Chapter 9 for more information on filter application and maintenance.)

VOLATILE ORGANIC COMPOUNDS (VOCS)

Many VOCs defy individual treatment. Increased ventilation is frequently used to reduce concentrations. Janssen, chairman of the ASHRAE committee to revise the old ventilation standard, 62-1981, indicated an effort to control these potentially harmful gases prompted a move from 5 cfm/person to 15 and 20 cfm/person.

Selective purchasing of construction materials, furnishings, maintenance and operational materials can avoid or reduce levels of VOC emissions. Metal shelving, equipment, appliances with a powder coated finish, rather than the conventional painted surface, offer essentially no off-gassing.

Materials, such as custodial cleaning materials, should be stored in well-ventilated places away from occupied areas.

"Bake-out" procedures (high temperatures to encourage off-gassing followed by air purge) have been found effective in reducing VOCs associated with new construction and refurbishments in some studies. Other research has found no benefit in using this procedure.

Time of use can be a key factor. Floor wax, for example, has a very high initial emission factor, which is followed by low-level steady state emissions. EPA research has shown floor wax emissions drop from 10,000 $\mu g/cm^2$ to about 500 $\mu g/cm^2$ in about an hour, and fall below 10 $\mu g/cm^2$ in 10 hours.

Direct exhaust or additional ventilation should be used for activities known to have high VOC emissions, such as spray painting. These activities should be conducted away from occupied zones whenever possible.

Table 6-2 offers a brief summary of primary control techniques for the major contaminants.

TABLE 6-2. SUMMARY OF CONTROL TECHNIQUES BY CONTAMINANTS/SOURCES

CONTAMINANT	SOURCES	CONTROL TECHNIQUES
Asbestos	Furnace, pipe, wall ceiling insulation fireproofing, acoustical and floor tiles	1. Enclose; shield 2. Encapsulate; seal 3. Remove 4. Label ACM 5. Use precautions against breathing when disturbed
Bioaerosols	Wet insulation, carpet, ceiling tile, wall coverings, furniture, air conditioners, dehumidifiers, cooling towers, drip pans and cooling coils. People, pets, plants, insects, and soil	Use effective filters. Check and clean any areas with standing water. Be sure condensate pans drain and are clean. Treat with algicides. Maintain humidifiers and dehumidifiers. Check & clean duct linings. Keep surfaces clean.
Carbon Monoxide (CO)	Vehicle exhaust, esp. attached garages; unvented kerosene heaters and gas appliances; tobacco smoke; malfunctioning furnaces	Check and repair furnaces, flues, heat exchangers, etc. for leaks. Use only vented combustion appliances. Be sure exhaust from garage does not enter air intake.
Combustion Products (NO_x, and CO)	Incomplete combustion process	Use vented appliances and heaters. Avoid air from loading docks and garages entering air intake. Check HVAC for leaks regularly, repair promptly. Use plants as air cleaners. Filter particulates.
Environmental Tobacco Smoke (ETS)	Passive smoking; sidestream and mainstream smoke	Eliminate smoking; confine smokers to designated areas; or isolate smokers with direct outside exhaust. Increase ventilation. Filter contaminants.

TABLE 6-2. SUMMARY OF CONTROL TECHNIQUES BY CANTAMINANTS/SOURCES continued

CONTAMINANT	SOURCES	CONTROL TECHNIQUES
Formaldehyde (HCHO)	Building products; i.e., paneling,, materials with lower particleboard, plywood urea-formaldehyde insulation as well as fabrics and furnishings	Selective purchasing of materials with lower formaldehyde emissions. Barrier coatings and sealants. New construction commissioning.
Radon	Soil around basements and slab on grade	Ventilate crawl spaces. Ventilate sub-slab. Seal cracks, holes, around drain pipes. Positive pressure in tight basements. NOTE: Increased ventilation does not necessarily reduce radon levels.
Volatile Organic Compounds (VOCs)	Solvents in adhesives, cleaning agents, paints, fabrics, tobacco smoke, linoleum, pesticides, gasoline, photocopying materials, refrigerants, building material	Avoid use of solvents and pesticides indoors. If done, employ time of use and excessive ventilations. Localized exhaust near source when feasible. Selective purchasing. Increase ventilation.

Chapter 7

An Ounce Of Prevention: Operations and Maintenance

The energy efficient building is less forgiving of poor maintenance. Problems that were once "blown away" in the old leaky buildings are now amplified and remain to haunt us.

To save energy, leaks in the building were plugged to reduce infiltration and exfiltration. Infiltration represented uncontrolled outside air, which caused drafts and discomfort. Efforts to "tighten" a building and reduce infiltration not only cut the costs of conditioning this unwanted air, but added to occupant comfort. The "tight" building, however, has lessened an important incidental means of diluting indoor pollution.

In another energy cost cutting effort, natural ventilation was reduced and mechanical systems were changed to recycle more of the already conditioned air. Unless the recirculated air was filtered and cleaned sufficiently, the pollutants just kept coming around and around like old suitcases on a baggage claim carousel. As each new "flight" arrived, the baggage piled up. Microorganisms, provided happy breeding grounds by poor "ground control," were trapped and recycled in the mechanical ventilation system. As they recycled in the tight building, the concentration of pollutants increased.

Poor maintenance, particularly in the mechanical system, is repeatedly cited as a major cause of indoor air pollution.

INHERITED "ENERGY CRISIS" PROBLEMS

Our energy problems of the 1970s proved to be a double-edged sword, cutting at maintenance both ways. The energy efficient building has not only made poor maintenance embarrassingly obvious, but climbing utility bills continue to be a major reason for cuts in maintenance budgets. This is particularly true in public institutions where budgets tend to be more rigid.

In 1979, the American Association of School Administrators asked their membership, the nation's school superintendents, where they were getting the money to pay higher utility bills. 68 percent responded that the primary source of the needed energy dollars was taken from operations and maintenance (O&M) budgets. And in that O&M "pocket," personnel costs were the hardest hit. Fewer positions, along with traditionally lower salaries, can cripple the maintenance efforts of dedicated plant engineers and facility managers.

Ironically, poor maintenance increases utility demands; prompting more cuts in the maintenance. This vicious downward cycle has left its mark on the quality of air we breathe.

In addition to cuts in manpower, a slow-down in equipment replacement as well as low first-cost decisions have compromised the facility manager's ability to maintain a productive environment. By 1983, the "Maintenance Gap" in our nation's public schools was $25 billion. A 1988 study of college and university facility needs determined that the cost of capital renewal and deferred maintenance had reached $70 billion.

Responding to the need to cut energy costs, mechanical systems have become more complex. Modern air conditioning and ventilation systems developed seemingly unlimited possibilities of combining prefabricated components into complex systems. Mixed air chambers, supply/exhaust air systems, combined elements, heaters, filters and mixers, exhaust/outside air heat exchangers, electrostatic filters and reserve elements for subsequent extensions requires a level of sophisticated maintenance not always available. Frequently, operations and maintenance (O&M) training has simply not kept pace. The absence of trained personnel to operate and maintain more sophisticated equipment as designed compounds the problem.

THE HIGH COST OF NEGLECTED MAINTENANCE

Many problems associated with indoor air quality are directly or indirectly related to operations and maintenance. Economies in the O&M area, which are apt to reduce the quality of indoor air, prove to be very costly in the long run. Rather than incurring modest O&M costs, organizations are increasingly laying themselves open for higher IAQ costs. Costs that could, and should, have been prevented.

Once the damage is done, attempts to identify contaminants are generally costly and often unreliable. Measurement procedures for many contaminants are still in the embryonic stages and talented professionals to interpret the results are in short supply. The multifactored, multidisciplinary nature of indoor air problems can further frustrate contaminant identification.

On the other hand, many sources and the means to control those sources are well known and many are traceable to maintenance practices. Since the costs to treat IAQ problems can far surpass the price of preventive maintenance, a new emphasis must be placed on the role of maintenance in assuring a safe, productive environment.

When O&M work, relegated to a lower priority, results in poor indoor air quality, there are apt to be social/political costs as well as remedial costs. Management may have a difficult time overcoming resentment when occupants realize they have been feeling miserable for days, weeks, even months because simple maintenance procedures were ignored. Economic losses associated with absenteeism and productivity can pale when compared to the long run impact of negative employee attitudes towards a management that didn't care enough to provide a healthy environment. To cap it all, the employees tend to expect a rather exotic solution to SBS problems and may not readily accept the idea that a simple maintenance action will now solve their problem.

Poor maintenance can, of course, be viewed by occupants as poor management ... even negligent management; perhaps legally negligent management. The administrative and financial burdens associated with potential lawsuits further tip the balance in favor of effective operations and maintenance.

OPERATION & MAINTENANCE: THE KEY IAQ INGREDIENT

Few people would care to take a ride on an airplane that they knew had not been properly maintained; yet, we regularly subject ourselves and others to unnecessary health concerns due to inadequate building maintenance. It is simply impossible to be assured of quality indoor air in a modern facility without a quality maintenance program.

Study after study lists insufficient maintenance as a primary cause of indoor environmental problems. To list only a few:

Writing in 1989, Janczewski and Yareb observed, "We have found that the majority of causes and solutions to indoor air quality (IAQ) center on the operation and maintenance of mechanical systems."

In his 1988 study of 21 buildings Morey found, "The most frequent cause of microbiological contamination in buildings was the inadequacy or absence of preventive maintenance," and concluded, "Reservoirs of any humidifier if improperly maintained can become a microbial amplifier."

Summarizing the Honeywell Indoor Air Diagnostic team's experiences, Mahoney noted that the frequency of causes in problem buildings they had identified could be loosely grouped into three areas; (1) maintenance and operations, (2) system design, and (3) contaminant load changes. The greatest of these, he observed, is poor maintenance. More specifically, Mahoney identified three areas where poor maintenance was most apparent: the lack of, or inappropriate, mechanical maintenance; misapplied energy conservation techniques; and changed loads.

Quality maintenance requires an effective facility management program. Reactive maintenance that simply responds to problems is sporadic and out of control. The facility management approach has four essential elements:

1. Top management support and commitment to the facility program, especially to operations and maintenance;
2. Planned O&M procedures; including a sound PM program;
3. Qualified personnel; and
4. Fiscal resources necessary to provide the manpower and materials required.

Operating the building and its equipment to meet environmental needs requires good maintenance and conscientious custodial practices carried out according to manufacturers' schedules and directions.

Building managers frequently experience turnover in operations and maintenance staff. The prevailing need to train new O&M staff means many facilities, especially public institutions, are usually playing "catch up" and are ripe for indoor air pollution problems. In an attempt to respond to indoor air problems caused by poor maintenance, the expedient, and usually costly, answer has been to increase outside air intake. This knee jerk reaction, relying totally on ventilation, is rarely the best solution. Dilution can mask the existence of potentially dangerous materials that may have long term hazardous effects. Settling for chronic low level exposure will leave us far short of what can be done to improve the quality of indoor air. Removing contaminants at the source, whenever feasible, is safer and more cost-effective than wasting energy using unnecessary ventilation.

The damage, disruption and human concerns related to each IAQ episode far outweigh the costs associated with an effective program. A relative small investment that provides training in identification, protocol and corrective maintenance procedures can save big headaches and costs later. As discussed in Chapter 3, the best management approach is a proactive approach, and "proactive" translated into O&M language is preventive maintenance.

PREVENTIVE MAINTENANCE FOR QUALITY INDOOR AIR

As every facility manager knows, a well-planned preventive maintenance (PM) program can prevent small deficiencies from blossoming into major, costly breakdowns, repairs and replacements. A simple thing like the routine oiling of bearings in a fan can prevent a loss of make-up air needed to assure quality indoor air.

Preventive maintenance has always been economically defendable, for it can:

- reduce unplanned service calls and the associated loss of man hours;
- reduce the number of equipment breakdowns;
- cut down the need for replacement materials and parts;
- reduce operating costs;
- create a more effective work environment for maintenance personnel;
- lengthen equipment life; and
- increase energy savings.

Responding to IAQ concerns, we can now add:

- fewer IAQ complaints and the administrative time required to resolve them;
- a more productive environment; and
- a decrease in absenteeism.

Clearly, the argument for preventive maintenance has become even stronger. The threat of indoor air problems has made preventive maintenance a critical function of effective management.

Setting IAQ implicated PM priorities as well as determining their timing and sequencing should consider known contaminant sources, building use schedules and equipment needs. For instance, the majority of maintenance-based IAQ problems are centered in the mechanical system; so this area deserves greater emphasis. Also, PM timing recognizes that pest control, painting and other actions involving the use of volatile organic chemicals should be done while few, if any, occupants are in the area.

Whether tracked manually or by computer, PM procedures should describe the actions to be taken and the frequency required. The dates and times (start and finish) of completed actions, the installation of any new parts, and the person who did the work should be a matter of record.

PM planning can incorporate IAQ concerns in a number of ways. PM procedures may be described in relation to; (A) probable sources of specific contaminants and the preventive measures needed, (B) needs related to specific equipment/systems, (C) services required by a particular component common to several pieces of equipment, (D) the facility, or (E) in relation to energy efficiency measures.

A. Preventive Maintenance: Specific Contaminants

Table 7-1, taken from the Anne Arundel County Public Schools' (AACPS) manual, Indoor Air Quality Management Program, lists the sources and nature of contamination for a group of pollutants: bioaerosols. The AACPS approach was developed within the context of a comprehensive IAQ program.

Using the table helps develop awareness of sources of a particular contaminant and implies, or directs, that some measures be taken in relation to those sources. It is a list that could also serve to remediate a problem

TABLE 7-1. PRESUMPTIVE SOURCES OF BIOAEROSOLS IN THE INDOOR ENVIRONMENT

SOURCE	NATURE OF CONTAMINATION
Cooling coil section of air handling unit	Amplification of microorganisms may occur on wetted surfaces of cooling coils and in drain pans.
Humidifier containing reservoirs of stagnant water	Organic dusts and debris are scrubbed from the air stream; microorganisms may amplify in water reservoir and on wetted surfaces of device.
Steam humidifier	Condensed water from improperly trapped devices may serve as a niche for amplification of microorganisms.
Air washer; water spray system in air handling unit	Microorganisms grow on wetted mechanical surfaces, in water reservoirs, and on porous substrates associated with these devices; see comments for 1 and 2 .
Fan coil units; induction units	These devices may serve as reservoirs for microbial contaminants; check for accumulated dust debris; see 1 for comments.
Filters in air handling and fan coil units	Where maintenance is poor, filters may function as reservoirs for microbial contaminants; amplification can occur if humidty is excessive or if filter is wet.
Porous man-made insulation in ventilation sytem	Dirt and debris are trapped in porous areas; can become microbial reservoir; if insulation is wet microbial amplification can occur.
Outdoor air	Outdoor bioaerosol are the ultimate souce of most indoor contamination.
Wet materials and furnishings	Ceiling tiles, carpet especially with natural fiber padding, wicker furniture, upholstery, etc., may function as microbial amplification sites.
Relative humidity greater than 70% in indoor environment	The equilibrium moisture content in organic dusts and substrates may rise to the point where it may support the amplification of microorganisms.

Source: Anne Arundel County Public Schools

when the known cause has been narrowed to bioaerosols. Just as important, this list taken within the PM context can help avoid the necessity to use the more agonizing and costly identification and mitigation processes. Table 7-1 can also serve as a model for incorporating other known sources/controls, such as those listed in Table 6-2 at the end of the preceding chapter, into a PM program.

B. Preventive Maintenance: Specific Equipment/Systems

Table 7-2 takes a different approach and lists traditional PM tasks that have IAQ implications for specific items of equipment within the air handling system. It does so without reference to specific contaminants. Table 7-2 illustrates the way the traditional PM approach can be used to satisfy IAQ needs with a subtle shift in emphasis. In this way, IAQ can be woven into an existing manual, or computer-based PM program, without a major change in operations.

Until the O&M staff has had IAQ training, this may prove to be a more expedient approach, as the PM staff can do the work without acquiring an understanding of the underlying IAQ needs. This approach, however, should never be regarded as a substitute for O&M training. Nothing takes the place of the staff understanding why they are taking certain actions, an awareness of when things "just aren't right," or that a particular action did not bring the expected results.

Table 7-2 is an illustration of how a PM program can be modified with reference to a specific equipment/system. HVAC maintenance is treated in depth in Chapter 9.

C. Preventive Maintenance: By Specific Components

Condensation pans are an excellent example of a piece of equipment that needs attention wherever found. They are perennial sources of indoor air problems if not properly maintained. Unattended condensate pans offer a dark, moist environment that is ideal for biological growth. They should be periodically cleaned and checked to be sure they are draining properly. Whenever a non-draining condensate pan is found, it should be repaired to operate properly and treated with an algicide according to product labeling.

TABLE 7-2. PREVENTIVE MAINTENANCE TASKS: AIR HANDLING SYSTEM

		FREQUENCY				
ITEM	PROCEDURE	WKLY	MNTHLY	QRTRLY	SEMI ANNUAL	ANNUAL
AIR HANDLING SYSTEM						
HEAT/ COOL/ VENT	CHECK FILTERS; CLEAN IF NEEDED; CHECK CONDENSATE PAN DRAINS	•				
	CLEAN OR REPLACE FILTERS	•				
	CLEAN COILS & SPRAY WITH DISIN-FECTANT	•				
	CLEAN & FLUSH CONDENSATE PANS WITH DISINFECTANT AS NEEDED			•		
	CHECK & CLEAN AIR INTAKES			•		
	CHECK OVERALL DUCT SYSTEM FOR CLEANLINESS, LEAKS, MOIS-TURE, COLLAPSE			•		
	CLEAN HEATING/COOLING COILS		•			
	CHECK & CLEAN FAN/BLOWER BLADES OF DIRT & TRASH BUILDUP		•			
	CALIBRATE CONTROL SYSTEM				•	
RETURN AIR FAN	CHECK & CLEAN AIR INTAKES		•			
	CHECK OVERALL DUCT SYSTEM FOR CLEANLINESS, LEAKS, MOIS-TURE, COLLAPSE		•			
	CLEAN FAN & BLOWER BLADES	•				
CABINET HEATERS	CLEAN & CHECK FILTERS AS NEEDED	•				
	CLEAN OR REPLACE FILTERS		•			
	CLEAN COIL AND SPRAY WITH DIS-INFECTANT		•			
FAN COOL UNITS	CLEAN & FLUSH CONDENSATE PANS WITH DISINFECTANT AS NEEDED			•		

TABLE 7-2. PREVENTIVE MAINTENANCE TASKS: AIR HANDLING SYSTEM cont.

		FREQUENCY				
ITEM	PROCEDURE	WKLY	MNTHLY	QRTRLY	SEMI ANNUAL	ANNUAL
FAN COOL UNITS, cont.	CHECK & CLEAN FILTERS AS NEEDED	•				
	CHECK & CLEAN COIL WITH SPRAY AND DISINFECTANT			•		
CEILING FANS	CHECK & CLEAN FAN BLADES			•		
UNIVENT UNITS	CHECK, CLEAN OR REPLACE FILTERS AS NEEDED		•			
	CHECK & CLEAN COIL AS NEEDED			•		
	CLEAN COILS & SPRAY WITH DISINFECTANT			•		
	CLEAN FAN BLADES & HOUSING				•	
COOLER	CHECK FILTER. CLEAN AS NEEDED	•				
	CHECK GENERAL FILTER CONDITIONS			•		
	DRAIN UNIT, CLEAN & FLUSH SUMP & STRAINERS			•		
	SECURE UNIT FOR WINTER					•
HUMIDIFIER	CHECK FILTERS. CLEAN SUMP AND STRAINERS			•		
	LUBRICATE FANS & MOTOR BEARINGS				•	
INCREMENTAL UNIT	CLEAN OR REPLACE FILTERS		•			
	CLEAN COILS AND SPRAY WITH DISINFECTANT		•			

TABLE 7-2. PREVENTIVE MAINTENANCE TASKS: AIR HANDLING SYSTEM cont.

ITEM	PROCEDURE	FREQUENCY WKLY	MNTHLY	QRTRLY	SEMI ANNUAL	ANNUAL
INCREMENTAL UNIT, cont.	CHECK CONDENSATE PAN,, CLEAN PAN & DRAINS, FLUSH WITH DISINFECTANT AS NEEDED		•			
	CHECK & CLEAN COIL			•		
ROOF TOP UNITS						
HEAT/ COOL/ VENT	CLEAN OR REPLACE FILTERS	•				
	CHECK OPERATION OF ROLL FILTER		•			
	CLEAN COILS & SPRAY WITH DISINFECTANT		•			
	CLEAN & FLUSH CONDENSATE PANS WITH DISINFECTANT AS NEEDED		•			
	CLEAN ALL COILS	•				
	CHECK CONTROL OPERATION				•	

Adapted from Indoor Air Quality Management Program, AACPS

D. Preventive Maintenance: The Facility

Since the HVAC system is frequently cited as a source of contaminants and as the mechanism dedicated to providing a comfortable environment, IAQ PM attention is rightfully focused on this area. There are, however, critical concerns connected with the building envelope and non-mechanical operations. Cracks in the foundation or openings around drain pipes, for example, allow radon to enter facilities. Bioaerosols grow in wet insulation or carpeting. Drains in laboratories are frequently overlooked as contaminant sources. Biological organisms find the sediment in drain traps an ideal location to gather and grow. Any location where water or sediment can stand in the plumbing should be kept clean and checked periodically.

A review of contaminant sources/control will suggest a series of building envelope and non-mechanical inspection PM measures that should be taken on a regular basis.

E. Preventive Maintenance: Energy Efficiency and IAQ

Blaming the energy efficient building is all to frequently used as a "cop out" to cover poor maintenance or to avoid a diligent search for the real problem.

Increasing energy efficiency as we improve IAQ has merit and is feasible. Rather than layering IAQ requirements over energy needs, they should be regarded as integral parts of the product building operators need to deliver: a comfortable, productive environment. Rather than competing goals, IAQ and energy efficiency can and should share a commonality of purpose.

Not all efforts to improve indoor air quality need to be done at energy expense. Some measures are energy neutral, including:

- checking duct linings, which are contaminant breeding grounds, especially in areas of high humidity, correcting any problems;
- exercising care that steam for humidification has no contact with boiler additives;

- removing partitions that impede critical air flow, or inserting grills or in-the-wall fans in the partitions;
- exercising great caution in the use and storage of chemicals in operations, maintenance, pest control, kitchens, etc. and using less toxic chemicals wherever possible;
- sealing cracks and openings around basement drains and openings, especially in radon affected areas; and/or
- positioning air intake grills to avoid the reentry of fumes from the building's own exhausts or other avoidable outside contaminants, such as fumes from delivery alleys and loading docks.

Many factors that have a negative impact on energy efficiency also have an adverse effect on IAQ. Correcting such situations can improve energy usage and enhance the quality of the air. Measures include:

- poor maintenance of pulleys, belts, bearings, heating and cooling coils and other mechanical systems can increase resistance causing a decrease in air supply. Good overall maintenance improves both energy efficiency and IAQ;
- water damaged insulation, ceiling tiles, rugs and internal walls support biological growth. Wet materials nullify insulating properties. Replacement increases energy efficiency while removing sources of biological contaminants;
- restricting uncontrolled infiltration improves comfort and reduces the air conditioning (heating and cooling) load on the equipment;
- leaks in terminal boxes and valves reduce temperature control, cause occupant discomfort and waste energy;
- humidity control can reduce the likelihood of mold while contributing to comfort and energy efficiency;
- malfunctioning, or inappropriately set, controls for outside air can bring in too much or too little air. Admitting too much outside air can actually burden the heating/cooling system beyond its capacity to respond, which detracts from occupant comfort. Running a

mechanical cooling system when outside air can supply cooling needs is wasteful and works against occupant comfort and health;

- shutting down ventilation systems during unoccupied hours does not, in most instances, affect the quality of indoor air as long as they are turned back on in time to create a comfortable healthy environment prior to occupant arrival; and
- frequent causes of combustion contamination are defective central heating systems in which the exhaust is not vented properly, or there are cracks or leaks in equipment. Defective or poorly maintained systems are less efficient and, at the same time, pollutant sources.

Clearly energy efficiency and indoor air quality are not only compatible components of the indoor environment, but many enhancement measures are mutually advantageous.

Special PM Activities

In almost every facility there are special PM needs with IAQ implications that warrant careful attention. Concerns related to a university's electrical transformers and capacitors, as discussed below, is an excellent example.

Federal law required that the university inventory its transformers and capacitors by October 1, 1990. The law also stipulated that the equipment be labeled and that location information be supplied to the local fire department.

In complying with these federal requirements, the university recognized that the polychlorinated biphenyls (PCBs) contained in this equipment, particularly if a fire were involved, posed a serious threat. The university, therefore, took a number of other steps as well. The director of physical plant included provisions in the PM program to dilute PCB concentrations in transformers to below 50 ppm and, wherever possible, to dispose of the liquid or metal casing. A scheduled replacement program was initiated and, until such time as the PCB transformers could be replaced, steps were taken to block ventilation and floor drains to prevent toxic combustion products from reaching occupied areas in the case of a fire.

The O&M emergency preparedness plan also included specific provisions for handling PCB fires or spills. A log book, also required by law, was set up to track inspections of PCB equipment and to record the approved disposal of the material.

PUTTING IT ALL TOGETHER

The approaches and specific concerns discussed above offer a number of ways to be sure IAQ needs become an integral part of a PM program. An elementary knowledge of contaminants and a little common sense in conjunction with accepted O&M procedures, will make any of these approaches workable. Contaminant sources discussed in Chapter 4 and controls and treatments presented in Chapter 6 provide a sound basis for identifying existing PM tasks that need greater IAQ emphasis.

By comparing traditional PM procedures with IAQ prevention measures, it will become evident that incorporating IAQ into PM program adds very little to the work load. In fact, the comparison procedure only serves to underscore the importance of preventive maintenance in maintaining a healthy productive indoor environment.

Whether IAQ concerns are addressed through a new or existing PM program, or some type of comprehensive maintenance effort, maintenance items that must be included can be summarized into a rather global items check list:

- [✓] change filters in all HVAC equipment regularly;
- [✓] clean coil and drain pans on a regularly scheduled basis;
- [✓] perform maintenance on mechanical system on regular schedule;
- [✓] maintain and calibrate controls on scheduled basis;
- [✓] regularly inspect and replace as necessary air duct liners;
- [✓] specify cleaning agents, procedures and schedule particularly for use in occupied spaces; and
- [✓] regularly inspect and repair all building envelope leaks.

POOR MAINTENANCE BY DESIGN

The annals of facility managers' lore are full of stories about air filters under a stage that never got changed, fan coil and induction units so difficult or time consuming to disassemble that they are never cleaned, equipment crowded against walls so that parts that require servicing can't be reached, etc., etc. The failure on the part of designers or builders to recognize maintenance needs takes on new importance when efforts to maintain indoor air quality are considered. Unfortunately, some designers are constrained by the owner's desire to increase revenue producing space. As a consequence, mechanical space is limited and such problems often result. In summarizing his environmental studies of microbial contaminated buildings, Morey reported, "In 11 of 18 buildings, maintaining mechanical systems was difficult, if not impossible, because systems were improperly designed and inaccessible."

Morey listed several examples, of poorly designed units -- all of which have a familiar ring:

- Air-handling units were designed without access doors to the heat exchanger section.
- Small air-handling units and heat pumps were located in inaccessible spaces above ceiling tiles. In many units, it proved to be physically impossible to gain entry into mechanical components to clean coils, drain pans, and other internal surfaces.
- Air-handling units were located in rooms or plenums so confining that human access was all but impossible.

The indoor air problems related to design extend beyond maintenance concerns and are discussed more fully under controls and treatment.

Poor maintenance by design could be largely avoided if the "resident experts" were only consulted. Building owners are missing an excellent opportunity for valuable input if they don't encourage architects and engineers to talk with their facility managers, directors of maintenance, plant engineers, operations supervisors, etc. These people have the hands-on information about what works and what doesn't in existing facilities. They are also in a position to foresee difficulties in contemplated designs.

Poor maintenance by design is supported, and compounded, by some rather common fallacies, ignorance, or "easy-way-out" rationalizations.

MAINTENANCE FALLACIES THAT WORK AGAINST PRODUCTIVE ENVIRONMENTS

"Tight" Buildings Cause IAQ Problems

This is a great excuse to ignore leaks in the building envelope. The repeated use of the term "tight" building in referencing indoor air problems suggests it's a bad thing. It's not! A leaky building allows uncontrolled amounts of air into a facility. It creates drafts and local thermal discomfort. A leaky building permits unfiltered air to enter a facility. Tightening a building envelope permits greater ventilation control and cleaner air. Provided the outdoor air brought in by the mechanical system compensates for infiltration losses, the tight building is one step closer to controlling the quality of the indoor air we breathe.

The Set and Walk Away Fallacy

Or, the "I've taken care of that and now I can forget it..." misconception. Buildings are dynamic. Changes in needs, programs, functions and use are almost constant. Occupants continuously disrupt the status quo. Ask the plant engineer about his or her day-to-day experience, the story is consistent: occupants can't leave anything alone. If they can reach it, they'll change it.

The Self-Sufficient EMS Fallacy

Energy management systems (EMS) are sometimes viewed with trepidation by maintenance. Training is frequently inadequate, which leads to an assumption (perhaps a self-deluding one) that an EMS runs itself and doesn't need maintenance. EMS' major functions include the control of ventilation, heating and cooling. Poorly maintained EMS systems account for many indoor air problems.

"Everything Must Be Okay Because They Haven't Complained," Fallacy

This comes under the, "If it ain't broke, don't fix it" rubric. Unfortunately, the fallacy equates "ain't broke" with a lack of complaints. IAQ problems, such as radon, do exist without immediate symptoms or complaints.

This fallacy is further supported by overburdened under-trained staff. The problem is exemplified by the prevalence of staff who don't know how pneumatic controls work or how to maintain them. The danger lies in the critical role pneumatic controls can have in HVAC systems. As the HVAC "nervous system," they greatly influence the comfort and quality of indoor air.

Conversely, the problem is further exacerbated when the overburdened maintenance staff are called upon to "treat" complaints where no problem really exists.

New Construction/Design Fallacy

This fallacy has a number of aspects. First, there is an assumption by building operators that the design, because it is new, will satisfy health and safety concerns associated with the quality of indoor air. Unfortunately, many architects and engineers are not yet well informed on IAQ needs with regard to design or materials specifications. Many SBS problems are associated with new construction.

Second, since the equipment was just installed and since it is covered by warranty, there is a fallacious assumption that it must be operating at design; therefore, the O&M staff can just coast. In an organization, the senior custodians are often given the new building, which other custodians view as a plum assignment as there is "hardly anything to do."

"It's-Running-at-Design" Fallacy

This may be the biggest culprit of all, because equipment seldom, if ever, operates at design. First of all, the O&M staff will reduce the operation of any piece of equipment to their level of understanding. Bailing wire and chewing gum remedies are often preferable to admitting ignorance to the boss.

These problems are compounded in still another way. Day time staff may assume around-the-clock controls are in place. Or, they may never have been informed as to the purpose of a certain switch or lever. In some instances, operating personnel may be inclined to espouse adherence to standards and conditions they "ought" to follow; rather than own up to actual conditions. As a result, those who have done "midnight raids" on facilities frequently report that systems are not running as described.

Equipment age, inadequate or inappropriate maintenance and repairs all contribute to operations that stray from design specs. These problems are compounded by changing contaminant loads and equipment loads; so, even if equipment were operating at design once upon a time, chances are the current needs are no longer served.

These fallacies only begin to reveal how many maintenance problems are rooted in misunderstandings, misconceptions, and the lack of the "right" information. All of which points to the vital role training holds in maintaining a staff qualified to do the job. Bulwer Lytton once observed "The pen is mightier than the sword." When it comes to achieving a productive cost-effective environment, this sage comment might be paraphrased, "The pen is mightier than the screwdriver." To put it another way, a well-trained O&M staff is widely regarded as the most effective "tool" in implementing an IAQ program.

Specific training suggestions are included in the discussion of management procedures in Chapter 3. No chapter on maintenance, however, would be complete without stressing the vital role training plays in implementing a solid PM program. Or, the critical role a preventive maintenance program has in assuring quality indoor air.

Chapter 8

The Thermal Environment

Historically, the primary purpose of a heating, ventilating and air-conditioning (HVAC) system has been to provide the human organism a comfortable internal environment. Exactly what those environmental conditions should be takes on a little different perspective if we first consider the human organism as described so delightfully in engineering terms by H. E. Burroughs.

> Man is a complete, self-contained, totally enclosed power plant, available in a variety of sizes and reproducible in quantity. He is relatively long-lived, has major components in duplicate, and science is rapidly making strides toward solving the spare parts problem. He is waterproof, amphibious, operates on a wide variety of fuels; enjoys thermostatically controlled temperature, circulating fluid heat, range finders, sound and sight recording, audio and visual communication, and is equipped with automatic control called the brain.

What a marvelous machine! The sophisticated complexity of this machine seems to stress how very sensitive it must be to abnormal or hostile conditions. Not so, says Burroughs, "The truth of the matter is that the human body is relatively resistant and able to accommodate environmental conditions that would be totally disastrous for actual machines."

So what precisely does this machine need to be comfortable and efficient ...er, healthy?

COMFORT AND HEALTH

A comfortable internal environment for the human organism has traditionally been defined by temperature, humidity and "fresh" air. Health concerns stemming from contaminants in the indoor air are a relatively recent phenomena. We have assumed over time that the environment is healthy if the body is comfortable. And, conversely, in a healthy environment the body will naturally be comfortable.

As IAQ concerns have come under closer scrutiny, we have had to revise our thinking. Health and comfort are not synonymous. An indoor environment can be relatively healthy and still not be comfortable. Our human "machine" may accommodate a contaminated environment where other machines would fail, although it may be uncomfortable. Conversely, an indoor environment can be comfortable and still not be healthy. Occupants may be exposed to health hazards, such as radon or asbestos, and still feel perfectly comfortable. The earlier accepted relative humidity range, 20-80 percent, was based primarily on comfort concerns; the narrow range today, 40-60 percent, incorporates health considerations.

Comfort, or the lack of it, is affected by several personal and environmental factors. Measurable comfort parameters include;

temperature and localized thermal discomfort,
relative humidity, and
air stratification and air flow; i.e., ventilation.

When temperature and humidity exceed accepted comfort parameters, they can, by themselves, negatively impact on air quality and be detrimental to health. We now know that the comfort factors of temperature and humidity can also interact or influence contaminants that may affect health.

TEMPERATURE

Trying to provide a comfortable temperature for all building occupants prompts a slight paraphrasing of Lincoln's admonition, to read: ... but you can't please all the people all of the time. What constitutes a comfortable temperature for one individual may not be acceptable for another person in the proximate area. An individual's ambient temperature needs may vary with age, physiology, activity, clothing, air movement, humidity, uniformity of temperature, contribution of solar gain, number of people in an area and occupant preference.

Moreover, a comfortable temperature may also be a matter of perception. Facility maintenance directors regularly tell the story of removing the thermostat lock box cover in a room where the occupant perennially complained. As soon as the occupant could adjust the thermostat, so the story goes, the person quit complaining. What the occupant didn't know, of course, was that the thermostat had been disconnected. The psychological implications of the human organism's inability to control its own environment could also be a factor in such a situation.

Recommended temperatures are usually given in ranges for the heating and the cooling seasons. In addition, a particular portion of a building; e.g., gymnasium, or an occupant activity, may be specified. The most common gauge of comfort is the ambient temperature. Recommended temperatures typically range from 68° F to 74° F for the winter and from 73° F to 78° F during the summer. A precise temperature, in and of itself, does not contribute to, or detract from, indoor air quality as long as the occupants are relatively comfortable.

The ASHRAE Standard 55-1981, Thermal Environmental Conditions for Human Occupancy, recommends the use of operative temperatures. The standard considers the ambient air, the mean radiant temperature, clothing and the activity level in calculating desirable operative temperatures. Figure 8-1 displays the ranges of operative temperature and humidity for persons clothed in typical summer and winter clothing at light, mainly sedentary, activity.

Ambient temperatures are, of course, measured with a thermometer. Operative temperatures can be measured with an indoor climate analyzer or a thermal comfort meter. Figure 8-1, however, can be used to determine the desired temperature or relative humidity in an office. For example, if the ambient temperature is known, the appropriate relative humidity can be ascertained by reading the X axis and following the appropriate line up to the shaded comfort zone and reading across.

The most common complaint related to the thermal environment stems from localized thermal discomfort, where one part of the body is too hot or too cold. Typical occupant complaints are: "My back is too cold," or "My feet are too cold!" Localized thermal discomfort may be the result of air currents, radiant temperature asymmetry, ground temperatures that are too low or two high, or situations where the vertical temperature difference between head and feet is too great.

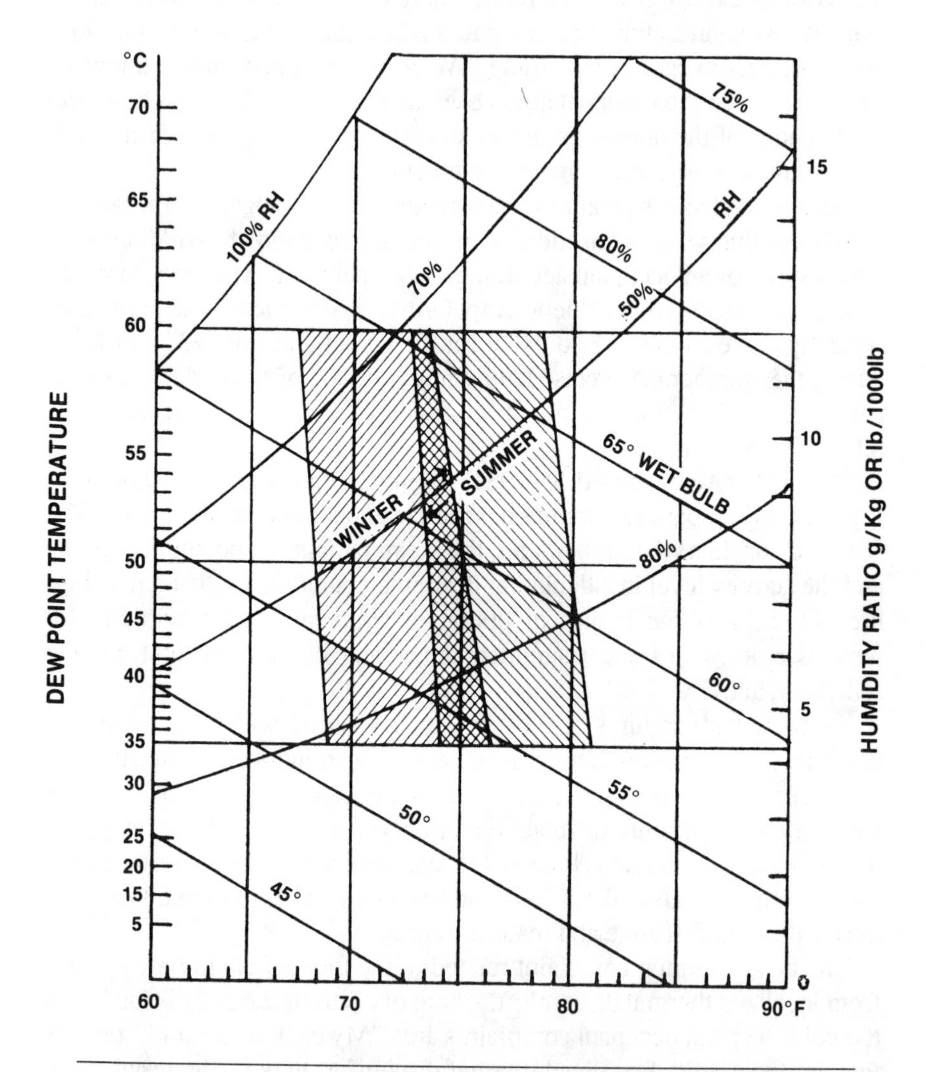

Reprinted by permission from ASHRAE Standard 55, copyright 1989, American Society of Heating, Refrigerating and Air-Conditioning Engineers, Atlanta, Georgia.

FIGURE 8-1. ACCEPTABLE RANGES OF OPERATIVE TEMPERATURES

There is increasing evidence that productivity and some health considerations are best met in the lower part of the accepted temperature ranges. According to recent studies, higher temperatures can affect mental acuity and are related to the appearance of some SBS symptoms. A reduction in mental work capacity has been observed when temperatures exceeded 24° C (75° F). Significant relationships between room temperatures above 22° C (72° F) and the existence of SBS symptoms have been found in several studies.

Inadequate and/or non-uniform heating/cooling can contribute to indoor air pollution and occupant sensitivity to contaminants.

High temperatures and humidity usually accelerates the off-gassing of volatile organic compounds, such as formaldehyde, from building materials and furnishings.

RELATIVE HUMIDITY

Relative humidity (RH) is the amount of water vapor in the air compared to what it can hold at a given temperature. The level of water vapor in the air affects the body's response to temperature. The RH level, therefore, affects the temperature ranges found to be acceptable to occupants. It is this relationship that primarily defines ASHRAE's operative temperature.

According to Proetz, an adult breathes approximately 500 cubic feet of air each day. The nose is required to bring the air to almost 98.6^{o} F and a relative humidity of 95 percent, or about 1.2 pounds of water for every 500 cu.ft. At 72° F and 40 percent RH, 500 cu.ft. holds about .25 pounds of water. Under those circumstances, the nose must supply .95 pounds, or about a pint of water every day. As the RH percent declines, the nose must make up progressively more moisture; e.g., at 20 percent, 1.08 pounds and at 12 percent, 1.125 pounds. As Proetz has observed, "It boils down to this: a pint of water is a lot of water for a small nose to turn out. In disease or old age it simply doesn't deliver; drainage stops and the germs take over."

HVAC designers have historically given little thought to the poor nose. They have focused on temperature and ventilation. But temperature control alone does not adequately treat all the physiological aspects required by occupants. Neither does ventilation.

Efforts to conserve energy since the mid-1970s have sometimes aggravated RH problems. Chilled water systems are designed to reduce the humidity from the incoming warm air. As chilled water temperatures were raised to reduce the energy consumption, the amount of humidity in the supply air was also increased. This higher humidity level made the air seem warmer to the occupants; and above 70 percent, more uncomfortable. In addition, the likelihood of bioaerosols increased.

Research has clearly documented the problems associated with air that is too dry or too moist. The relationship of RH to the prevalence of mites, fungi, viruses is clear. Airborne viruses, for example, survive best at lower, and higher, humidity levels. Influenza virus survives much better at lower humidities, poliomyelitis virus at higher humidities. The RH level, least hospitable to contaminants is at about 50 percent.

Morbidity rates for colds have a very seasonal pattern. The incidence of colds increase through the fall, peak in winter and then decline until May, where the rate tends to level off for the summer. About thirty years ago Lubart, writing in the New York State Journal of Medicine, described the relationship of RH and colds as follows:

> A dry nose and throat caused by artificial heating creates an indoor climate favorable to the cold-inciting agent. Dry nasal mucous is an excellent culture for the infective agent. Adequate moisture is necessary also for the proper function of the cilia, pharynx, larynx, trachea, and lungs. During the heating season the best method of prevention of colds is maintenance of a proper balance of humidity by means of such devices as mechanical humidifier. In summer, regulation of the humidity in air-conditioning is necessary to prevent summer colds.

Before 1960, Andrews found that 40-50 percent RH with temperatures of 68° F to 70° F produces the most healthy conditions for living and working areas and for recovery from diseases of the respiratory system. He observed that the 40-50 percent RH "reduces the incidence of respiratory infection and speeds recovery from the common cold." Studies throughout the 1960s and 1970s confirmed the value of proper humidity in the prevention, amelioration and relief of infections of the respiratory tract.

Despite years of amassing such data, HVAC designers still frequently fail to address humidification needs adequately.

If humidity prevents the drying and cracking of wood, leather, paper, etc., why has it taken so long to recognize that the human organism needs humidity, too?

There are two paramount reasons for this neglect. First, owners still view humidification/dehumidification as dispensable; therefore, when construction costs come in over budget, RH equipment is among the first to go. Second, HVAC systems are still being designed and installed using the same techniques developed over thirty years ago.

With the growing focus on indoor air quality, relative humidity is getting more attention and productivity/absenteeism considerations are apt to expand that attention. Productivity and performance has been shown to correlate with 30-60 percent RH. This is hardly surprising. Knowing that RH levels have been tied to the morbidity of colds is reason enough to suggest a relationship to productivity. Find a person with a miserable cold, who feels he or she is functioning at par or better, and you'll be looking at an atypical person. Studies by Ritzel, Sale, Gelperin and Green have all shown a statistically significant reduction in respiratory infections and absenteeism among occupants of buildings in which humidity is controlled adequately.

Humidity Control

Relative humidity can be maintained by providing humidification when RH is too low and dehumidification when water vapor in the air is too high.

Humidification. While air that is too dry presents a range of health problems, humidifiers that are inappropriate or poorly maintained can create even greater problems. Wherever possible, the humidification process should be limited to direct steam injection. Treatment of makeup water utilized in the steam-generating process should be done with care to avoid boiler additives entering the humidification steam.

Assuring proper moisture content in the air can be a money-making proposition. A methodology for calculating the economic benefits of humidifying an office building is offered by Berlin.

CALCULATION PROCEDURES: HUMIDIFICATION ECONOMIC BENEFITS

EXAMPLE: A 25,000 sq. ft. office building has an indoor temperature of 72° F. An electric humidifier to achieve 40% RH would require 34 kW to produce 103# water/hr. It would need to have a seasonal operating rate of 2,000 hrs. Electricity is $0.10/kWh without demand. Ventilation is 15 cfm/person; 225,000 CFH/O/A. 250 employees have an average salary of $30,000.

PROBLEM: If productivity losses due to a lack of humidity (comfort, respiratory problems, absenteeism, etc.) are 1 percent, what would be the net profit/loss of supplying humidity?

CALCULATIONS:

Grains of Moisture to Add: 3.20/cu.ft.
(Indoor: 72° F and 40% RH; 8.59 X 40% = 3.436)
(Outdoor: O° F, 50% RH; .48 X 50% = .24)

Humidification load: $\frac{\text{3.2 GRS X 225,000 CFH}}{\text{7,000 GRS/lb.}}$ = 102.86 lbs./hr

Operating cost: 34 kW X 2,000 hrs. X $.10/kWh = $6,800.00

Personnel costs: Avg salary $30,000 X 250 = $7,500,000
1% productivity loss = $75,000.00

ANNUAL SAVINGS: $75,000 - $6,800 = $68,200.00

In general, comfort systems are designed to accommodate the human "machine;" not special machines, such as data processing equipment. A room full of mainframe computers and control equipment may need a system that will respond with sufficient speed to maintain nearly constant temperature and humidity. In instances where building comfort systems are not adequate, computer-based precision air-conditioners that have a sensible heat ratio of 95-98 percent in contrast to the typical 65-70 percent may be needed. Computers throw off more heat and less moisture than people. Precision air-conditions that devote 95-98 percent to cooling and 3-5 percent to dehumidification better meet this specialized need.

Dehumidification. The removal of water vapor from the air is done by mechanical refrigeration or through a desiccant-based system. Mechanical refrigeration has a lower first cost and is more practical down to a certain level. In some instances, it may become necessary to dehumidify at a point where the refrigeration cooling surface would have to be below freezing to obtain the desired results. When the cooling surface drops below freezing, frost build up interferes with the efficiency of the cooling unit, the unit becomes more costly to run, and the flow of air may be impeded. Under such conditions, a desiccant-based system is warranted.

HUMIDITY AND ASHRAE 62-1989

Ironically, compliance with the new ASHRAE 62-1989, Ventilation for Acceptable Indoor Air Quality, can have a deleterious affect on the indoor air with regard to humidity. To the extent that the new ventilation requirements will increase the amount of outside air brought into a facility, it can have a detrimental affect on RH. In most climates, it will mean more humidity in the summer and drier air in the winter. Since most buildings are already burdened with a design that gave little consideration to controlling humidity, the new ASHRAE standard will intensify the problem.

The resulting severe decrease in indoor RH in most buildings during the winter will provoke more colds, respiratory and physiological problems associated with air that is too dry. Anticipating such a problem, the following case study describes the steps the Matthews Group took at their new Sussex Centre.

CASE STUDY: ASHRAE 62 COMPLIANCE AND HUMIDITY

NEW FACILITY: SUSSEX CENTRE
MISSISSAUGA, ONTARIO
Office complex/retail concourse

DEVELOPER: Matthews Commercial Construction

MECHANICAL CONSULTANT: Smylie and Crow Associates

Sussex Centre has two towers connected by a two story shopping concourse, which features a four theater cinema and a full service health club. Each tower is accented by a four story atrium. Even while ASHRAE 62 was in proposal stage, the Matthews Group decided the proposed 20 cfm/occupant was preferable. This outside air requires the HVAC system to introduce 30,000 cfm into each tower and 15,000 cfm into the concourse.

Recognizing that such a volume of outside air would cause a severe decrease in the indoor RH during the winter months, a decision was made to have individual Nortec steam humidifiers on each floor to help alleviate the problem. Rather than preheat fresh air at the source, the Sussex Centre uses a more economical approach. Fresh air is first mixed with warmer return air in the fan room to "free cool" the return air before supplying it to HVAC systems on each floor. Humidifiers are located in individual VAV compartments on each floor.

The free cooling saves energy that would have been used to condition the return air. The return air also preheats the fresh air supply through the winter months. During the summer, the fresh air/return air mix is treated by the building's cooling system before being distributed to each floor.

By selecting the packaged humidifiers over steam boilers many installation and operational problems were avoided. The problems associated with installing extensive steam and condensate return piping in the towers was avoided. Maintenance can be done one floor at a time, rather than shutting down the entire system.

In addition to the problems directly associated with respiratory problems and indirectly with bioaerosol growth, water vapor content in the air can significantly affect the release rate of many indoor contaminants and their concentrations in the air. Dry air fosters the breaking of fibers in carpets and other fabrics. Particles are suspended and recirculated for longer periods of time in dry air; thus, increasing the likelihood of inhalation. Higher RH levels increase off-gassing of VOCs, like formaldehyde.

VENTILATION

Ventilation is central to indoor air quality. A certain rate of air exchange is essential to a healthy indoor environment. With the advent of the energy efficient, or "tight," building, the drop in infiltration has made ventilation a more critical factor.

Ventilation is the most frequently cited IAQ problem and the most frequently prescribed control. Indoor air quality problems and their control are flip sides of the same coin. IAQ corrective measures respond to problems <u>within</u> ventilation systems and also <u>use</u> ventilation to dilute contaminants.

VENTILATION AS A CONTROL

Ventilation is an effective control procedure, but it is not a panacea. Ventilation as a control measure is a second choice; and a poor second choice. The control measure of choice is always elimination of the contaminant. Source control through elimination or substitution removes the contaminant threat completely. Dilution by ventilation only "waters" it down.

If a hazardous substance were dripping from a faucet in an occupied area, few would accept hosing it down each morning as the preferred treatment. The toxic substance would still be there, just diluted. For many contaminants, the long term health effects of chronic low level exposure is not known. Diluting by ventilation only masks the problem and leaves the potential for deleterious effects that could occur through chronic low level exposure. Settling on increased outside air and that alone must be avoided unless the contaminant is unknown and/or the source cannot be determined. Masking potentially dangerous materials that may have long term hazardous effects leaves us far short of what can be done to improve the quality of indoor air.

Much has been made of the National Institute for Occupational Safety and Health's (NIOSH) finding, "In 52% of our investigations, the building ventilation has been inadequate." This frequently quoted statistic has been taken as a mandate to turn up the fan. Inadequate ventilation does not equate to inadequate outdoor air.

It is important to go beyond the statistic and look at the "ventilation problems" NIOSH has encountered, which they describe as:

> ... not enough fresh air supplied to the office space; poor air distribution and mixing which causes stratification, draftiness, and pressure differences between office spaces; temperature and humidity extremes or fluctuations (sometimes caused by poor air distribution or faulty thermostats); and filtration problems caused by improper or no maintenance to the building ventilation system. In many cases, these ventilation problems are created or enhanced by certain energy conservation measures applied in the operation of the building ventilation. These include reducing or eliminating fresh outdoor air; reducing infiltration and exfiltration; lowering thermostats or economizer cycles in winter, raising them in summer; eliminating humidification or dehumidification systems; and early afternoon shut-down and late morning start-up of the ventilation system.

Inadequate outside air is only one of a long list of ventilation problems. Dilution is not the 52 percent solution. Ventilation, problems and solutions, is more than outside air.

NIOSH has pointed out that the 52 percent figure is based on soft data. To the extent, however, that it represents primary problems in the investigated buildings, the NIOSH findings also impart another critical piece of information that is typically overlooked: 48 percent of the problems found by NIOSH will NOT be solved by ventilation changes. NIOSH has the most extensive experience in indoor air problem investigations and a highly regarded protocol. They have determined that nearly half of the problems they have investigated are not related to ventilation. If the NIOSH data and problems identified by other investigation teams are considered collectively, it seems safe to surmise that a great many of our indoor air problems cannot be satisfied solely by increasing outdoor air intake.

Dilution through increased ventilation certainly offers a means of mitigating some contaminants. Increased air changes per hour may be beneficial when the contaminant or its source is unknown or can't be determined; or, it may serve as an intermediate step until action can be taken. Unique applications of ventilation; e.g. localized source control or sub-slab ventilation to control radon, are valuable control mechanisms.

VENTILATION LIMITATIONS

Ventilation does have its place. Dilution is a viable and valuable means of controlling pollution levels. But quantity does not equate to quality. Before we turn up the fan and run up the utility bill, we need to keep in mind that specifying the quantity of outside air is only one of many alternatives available and ignores many important factors, including:

(1) The quality of the outside air may be "bad;"

(2) More air may not help if the ventilation effectiveness is poor;

(3) There is no single ventilation rate that will assure adequate indoor air quality -- other factors frequently have a controlling effect;

(4) Increased ventilation can be a costly remedy. Costs to move and condition the air represent a significant portion of the utility bill. Increasing ventilation adds to the costs of heating/cooling. To the extent that increased ventilation requires the air handler motor to move more air, the power required to move that additional air is the cube of the flow rate; and

(5) Increased ventilation can bring with it serious humidification problems, particularly in the winter months, that may more than offset any gains achieved through dilution.

Ventilation Standards and Non-Occupant Related Pollutants

Ventilation requirements in cfm/occupant is historically based on body odor and smoking. Cfm/occupant does not address contaminant concentrations that are non-occupant related, such as formaldehyde or radon. As with many contaminants, it is the source strength that frequently determines the levels found indoors. For example, variation of radon concentrations and their entry rate has been shown to be much greater than changes in the ventilation rate. Ventilation rate is not the most important variable in radon concentrations in houses. Source strength determines levels found in the building.

Ventilation as a control for VOCs is still unclear. Walkinshaw et al. investigated VOC concentrations and ventilation rates in eight Canadian settings; an office, office/lab, office/library, two schools, two hospitals and a residence. Even with ventilation rates exceeding 15 cfm/occupant in all but the schools, the VOC concentrations met or exceeded levels associated with mucous membrane irritations and impaired ability to concentrate.

Hodgson et al. measuring VOCs in an office building and a school found that the apparent specific source strength for VOCs approximately doubled with a six-fold increase in ventilation rate. It was not determined what caused the increase in source strength.

On the other hand, Indoor Air Quality Update reported in May, 1989, "VOC concentrations tend to increase very quickly as ventilation decreases below a rate within the range of 0.6 to 1.2 ACH, depending on the source strengths and sinks. ... The key to economical IAQ control is to maintain ventilation just above the inflection point wherever possible."

It is unclear if the disparity in the research findings related to ventilation and VOCs can be attributed to the VOCs present, intervening or confounding variables. At this point, ventilation as a control for VOC contaminants is in doubt.

USING VENTILATION EFFECTIVELY

The list of ventilation problems found by NIOSH also serves to identify some ventilation solutions that do not require increased air intake. Temperature and humidity fluctuations and extremes as well as filtration problems should be on every ventilation checklist. Common errors in ventilation that frequently impact on the quality of the indoor air are:

(1) Capture, or "backdraft," hoods located too far from the source of contaminant generation.

(2) Exhaust stacks placed in close proximity to outside air intakes.

(3) The build up of dust on internal surfaces in exhaust systems, which causes increased resistance and reduced ventilation.

(4) Existing exhaust hoods not used effectively by employees.

(5) General exhaust ventilation too often relied on to control employee exposure to airborne contaminants.

(6) The use of flexible ducts in place of rigid ducts, requiring additional energy to move the necessary air volume.

(7) Centrifugal fans operating backwards, which still exhaust air at lower volumes.

(8) Belt driven fans with missing or slipping belts, missing or damaged ducts, which reduce ventilation.

(9) Inadequate make-up air, reducing the effectiveness of the exhaust system.

(10) Addition of hoods and ducts to pre-existing systems without adjustments for increased resistance losses.

(11) Placing walls or partitions in areas originally designed as "open space" without in-wall fans or grill work to help avoid dead air pockets.

(12) Adding copiers, computers, equipment or otherwise changing the functional use of the space without modifying the ventilation to accommodate the changes.

This list as well as earlier comments emphasize that design, maintenance and common sense are critical ingredients in operating a ventilation system and in using ventilation effectively as a control measure. Two aspects of ventilation that are often addressed in the IAQ literature warrant further discussion. These two concerns are the quality of the outside air brought into the facility and the effectiveness of the system in delivering the air to the occupants.

"Bad" Air

The quality of outdoor air surrounding a building is affected by external factors; e.g., garbage dumpsters, traffic, loading docks, etc. Designers need to evaluate more carefully the location of intake grills to minimize the outdoor pollution entering the HVAC system and to avoid the building's own exhaust reentering the building.

It is essential that engineers and architects recognize the effects of pressure gradients and air movement around buildings to project airflow particularly in city centers. Careful consideration to the way these factors influence air flow around a building can minimize the entry of contaminated air into occupied building spaces.

Bahnfleth and Govan reported on the sad circumstances related to a remodeling job when reentry was not adequately considered. A remodeled ground floor of an eight-story building was across the alley from a single-story office attached to a ten-story tower. A mushroom-type exhaust fan was installed in the exterior wall of the remodeled space to serve a steak and hamburger grill. Grease fumes, smoke and odors were exhausted into the alley at the ceiling of the first floor and taken inside through the air intake of the ground floor office.

As concrete evidence of the old adage that a little can go a long way, Bahnfleth and Govan observed that the occasional "whiffs" of cooking steak, which can whet one's appetite, has a negative impact when it becomes a continuous heavy dose of hydrocarbon fumes and odors. Increased absenteeism in one year alone accounted for a $32,000 loss. Lost productivity of those who remained on the job could not be measured but lost time was evident.

A centrifugal fan and a short length of duct work were installed to get the exhaust fumes up and away from street level solving the problem.

Ventilation Effectiveness

The flow of air within the facility and how well that air actually reaches occupants is a concern that increasingly haunts buildings owners and managers. Not only does the absence of ventilation effectiveness cause many SBS complaints, but it frustrates efforts to use ventilation as a control measure.

Most office HVAC systems are designed with the input and output grills near the ceiling. Such an arrangement can cause most of the air to by-pass the occupants as shown in the bottom illustration of Figure 8-2. This "short circuiting" creates air stagnation at occupant level and fosters complaints of stuffiness and other SBS symptoms. Increasing air intake to the ASHRAE

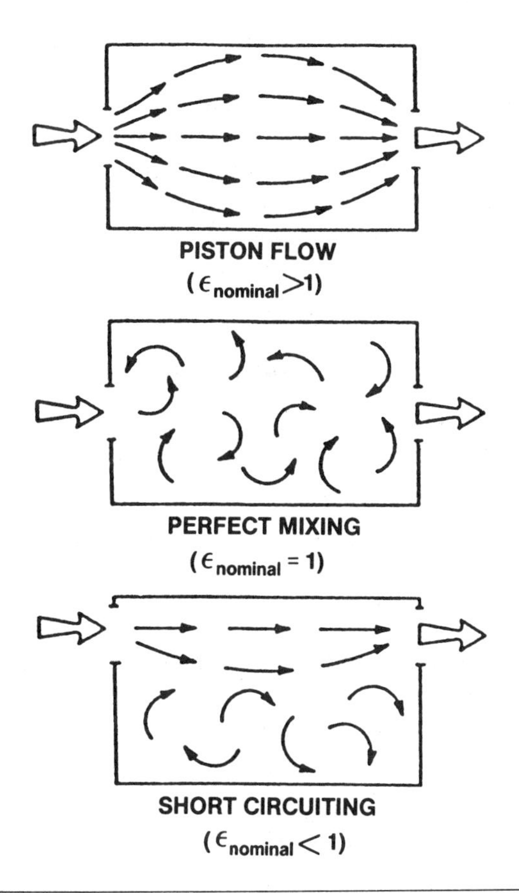

Source: U.S. Department of Energy

FIGURE 8-2. REPRESENTATIVE EXAMPLES OF VENTILATION AIR MOVEMENT

62 recommended level of 20 cfm may create a nice breeze across the ceiling level (at considerable cost), but it may not significantly reduce contaminant levels around the occupants. The greatest problem associated with ventilation as a control measure is the lack of ventilation effectiveness.

VENTILATION AND ENERGY EFFICIENCY

So much attention as been given to reductions in ventilation to cut energy consumption that ventilation and energy efficiency are generally regarded as adversaries. Studies coming out of Europe indicate that methods used to improve the quality of the inside climate also lead to the more efficient use of energy. For instance, Luoma-Juntunen of Finland have found that up to 20 percent of the energy used for ventilation can be saved by the more even distribution of fresh air between the rooms.

Fleming has listed a number of ways that ventilation can be used more efficiently in residences. His suggestions, which have application to other facilities, include:

> spot ventilation, using exhaust fans appropriate to the pollution source; use only while pollutants are being emitted:
>
> natural ventilation, opening the windows in mild weather;
>
> fans, which can cool and ventilate less expensively than air-conditioners; and
>
> heat exchangers, an air-to-air heat exchanger can transfer heat from the warm outgoing (inside) air so that 50 to 80 percent of the energy normally lost in the exhaust air is recovered.

ASHRAE predicts a 40 to 50 percent total building energy reduction is possible through the use of heat exchangers.

Fleming's recommendation to "open the windows" in homes may have limited carry over value to other facilities. In an era when "natural" is always assumed to be better and the inclination to throw open the windows is strong, the conclusions drawn by Feustel et al. take on added import. The researchers concluded,

> Our first set of conclusions from this comparison of ventilation strategies is based on the total airflow and indoor air quality resulting from each strategy. We found that all the mechanical ventilation strategies examined provided more uniform ventilation rates than natural ventilation and, thus, lower total airflow and potentially better indoor air quality.

Weighing natural and mechanical ventilation, Hedges et al. found 88 percent of those with air-conditioned offices reported too little ventilation compared with 60 percent of those in naturally ventilated offices. Since the findings were based on interviews, not measurements, it is not clear if occupant concerns were real or perceived. The Danish Town Hall Study reported, "The difference between mechanically and naturally ventilated buildings was not significant for this study." Thus, the question, "Is natural ventilation better?" remains open at this time.

The discrepancy in findings may be attributable to the quality of the outside air. Natural ventilation does not offer an opportunity to filter the incoming air. Futhermore, operable windows means occupants have control of the amount of outdoor air entering a facility. There is no quality control of indoor air with natural ventilation.

Chapter 9

At The Heart Of IAQ: HVAC

Since the heating, ventilating and air-conditioning (HVAC) system serves as the lungs of a facility, occupants depend on the HVAC system for comfort, ventilation, temperature, odor and humidity control. The system's effectiveness affects productivity, performance and, most importantly, health.

Unless the HVAC system is well designed and maintained, outdoor contaminants can travel into a building and jeopardize the quality of indoor air. Similarly, the system must filter, dilute and exhaust these pollutants or contaminant levels will increase. The HVAC system is a controller of indoor air quality; and, when not properly maintained, a polluter as well. Poorly maintained HVAC systems are a primary source of pollutants, such as biological growth.

Various studies have established that the HVAC system is responsible for 50 to 60 percent of building generated IAQ problems. The HVAC system is capable of resolving up to 80 percent of the indoor air problems. In other words, if a building is "sick," chances are the HVAC system is at fault, or can remedy the problem. The design, operation and maintenance of a HVAC system, therefore, is at the heart of a quality indoor air program. As such, it is deserving of special attention.

HVAC DESIGN

A well-designed HVAC system is the mainstay of a healthy building. Poor design is repeatedly cited as a major IAQ problem source. The Honeywell IAQ Diagnostic (IAQD) team has identified three related groups of IAQ problems. Design related problems are second only to operations and maintenance in contributing to indoor air pollution. More specifically, IAQD found the most design difficulties in:

(1) ventilation and distribution;

(2) inadequate filtration; and

(3) maintenance accessibility.

The design of many HVAC systems in use today originated in the temperature control era. In fact, system designers still tend to be concerned with temperature to the exclusion of humidity and indoor air quality. Owners, facing budget constraints, sometimes force this focus on designers.

During the past 20 years, HVAC design criteria have focused increasingly on energy efficiency and conservation. Variable air volume systems (VAV), in particular, found a home during this era as they are more energy efficient than constant volume systems. An almost concomitant goal during this same period was the effort to lower HVAC equipment's first cost.

Since the late 1970s, there has been a growing interest in HVAC design to enhance indoor air quality. In particular, the focus has gradually shifted to increasing the volume of outside air introduced, the effective distribution of the air, and air cleaning procedures. ASHRAE has spearheaded action in this area with 62-1989, <u>Ventilation for Acceptable Indoor Air Quality</u>, serving as the new construction and operating guidelines.

Design criteria for ventilation efficiency and effectiveness have been slow to emerge. Designers still tend to assume that air passing through diffusers will effectively reach occupants. Air flow patterns that "short circuit" the ventilation are still common place. Air cleaning technology for commercial and institutional buildings, for the most part, is still limited to treating contaminants in response to existing information, such as adaptations of industrial work place guidelines.

Energy demands of the 1970s and early 1980s spawned financing schemes for capital retrofits, which forced engineers to become increasingly accountable for predicted energy savings. The whole performance contracting industry today rests on the engineer's ability to predict the amount of energy a given measure will save. Similarly, as IAQ becomes a high stakes game of productivity and lawsuits, engineers and architects of record will be held increasingly responsible for the indoor air quality the HVAC system delivers. This responsibility is apt to prompt more professionals to stay involved beyond the design stage (which could easily be worth an annual fee to the owners). Greater accountability and prolonged involvement will inevitably enhance HVAC design.

INSPECTING THE HVAC SYSTEM

It is amazing what can be found in the air handling components of an HVAC system, if one only looks. Duct work is a disturbing example. If we could see the garbage heaps the air we breathe frequently pass through, we'd be appalled. Robertson describes just a few of HBI's findings:

> Excessive dirt accumulations are common in duct work, even in hospitals. Frequently dirt is built into the systems during construction since the ducts are installed long before the windows, etc. and construction dusts from the site, plus wood shavings, lunch packets, coke and beer cans, etc. find themselves brushed into the vents then "out of sight--out of mind."
>
> ...Dead insects, molds, fungi, dead birds and rodents are common. In 1984 we found two dead snakes in air supply ducts. We have also found rotting food, builders' rubble, rags, and newspapers. All of these contaminate the air we breathe. It is the dirt that encourages germs to breed -- germs which cause infections.

It is not surprising that buildings can make us sick! And that's only the duct work.

Many HVAC problems, such as obstructions in the outdoor air intake, can be caught by visual inspection. Just a little awareness and an educated eye can make a big difference. Other inspection procedures may require some simple measurements, such as the air flow at the diffusers, to see if the coil in a terminal box (terminal reheat) needs cleaning. HVAC inspections guidelines can be found in the "IAQ O&M Opportunities" listed under the specific components discussed later in the chapter. Chapter 5, Investigating Indoor Air Problems, also offers guidance.

HVAC OPERATIONS AND MAINTENANCE

It is virtually impossible to over-emphasize the importance of HVAC operations and maintenance (O&M). The inclination, is to first think of those components that come in direct contact with the air; e.g., filters and duct work. As important as that is, no part of the HVAC system can be ignored. The malfunction of any part of a HVAC system from worn bearings in a fan to dirty coil fins can affect the system's ability to provide quality indoor air.

There is much that HVAC technology can do to improve the quality of our air. Those opportunities seem more manageable if the component parts of a HVAC system are considered individually.

HVAC IAQ OPPORTUNITIES

The main purpose of mechanical ventilation is to provide a comfortable environment for the majority of its occupants. The IAQ focus makes sure it is a comfortable <u>and healthy</u> environment. To do this, the HVAC system provides many services including;

heating,
cooling,
outside air supply,
filtration, and
air distribution and diffusion.

A typical "H" system schematic, shown in Figure 9-1, provides a way to identify the key points in a system before some variations are considered.

The main components of a mechanical ventilation system, as illustrated in Figure 9-1 are standard. These components, their graphic symbols and descriptions may assist non-technical personnel in interpreting the schematics shown in this chapter. The symbols and descriptions follow Figure 9-1.

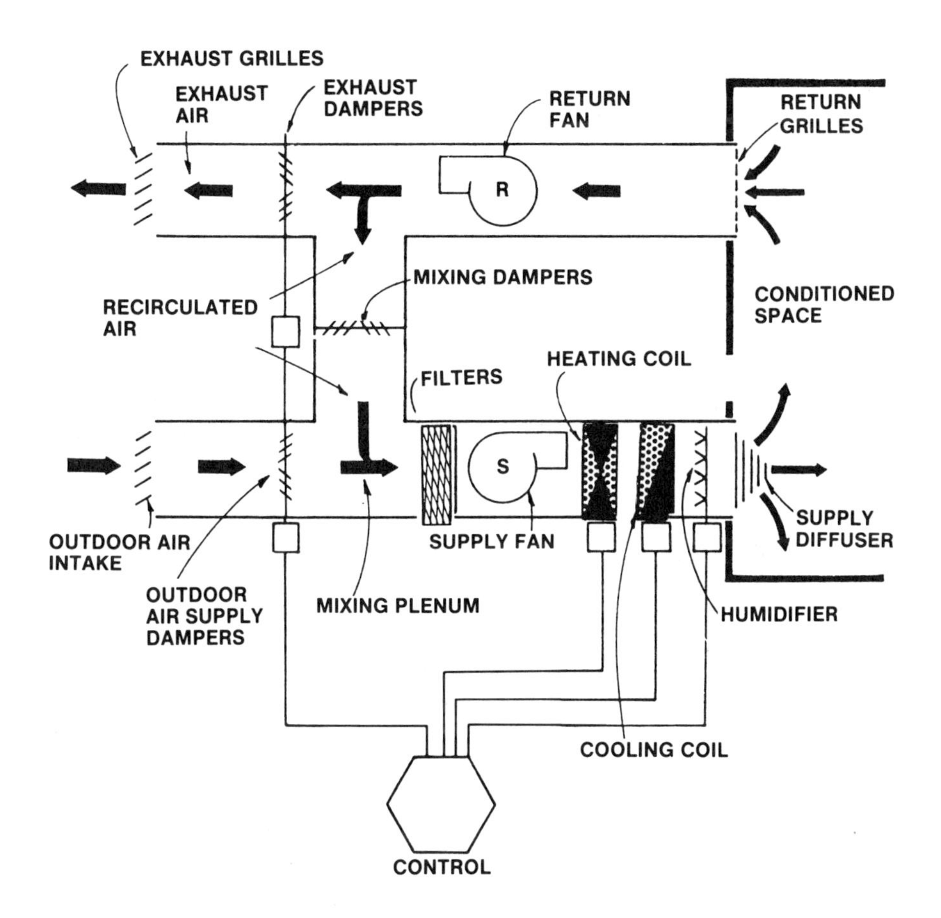

FIGURE 9-1. MECHANICAL VENTILATION SYSTEM

Controls 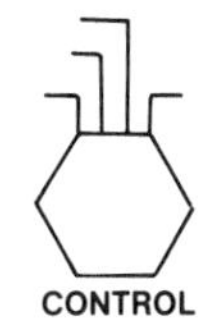

The operation of the components are governed by controls to achieve the desired comfort and air quality in as energy efficient mode as possible.

Cooling Coils 

Only air-conditioning systems have cooling coils. They are also used to dehumidify the air during the summer season.

Dampers		Exhaust, mixing, and outside dampers are designed to control the amount of air exhausted, admitted, recirculated and mixed. The dampers are linked and work in tandem.
Exhaust & Intake Grills/ Louvers		Mounted on external walls, the grills and louvers allow for the discharge of return air and intake of outdoor air. Intakes and exhausts have "bird screens" to keep birds, rodents, snakes, etc. out of the system.
Filters	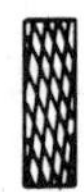	Filters screen particles and bioaerosols from the air to protect the equipment downstream and clean the air we breathe.
Heating Coils		One means of heating incoming air to adequate temperatures are heating coils.
Humidifier		Water vapor or steam is injected into the air to increase relative humidity. They are usually needed only in winter.
Return Air Grill		Air is removed from designated area through the return grill to ensure circulation.
Return Fan	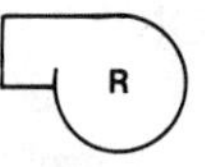	Return air is drawn from the rooms and pushed into the evacuation and mixing ducts by a return fan.
Supply Diffuser		Conditioned air at the terminal box is delivered to occupied areas by the supply diffuser to avoid drafts.

Supply Fan 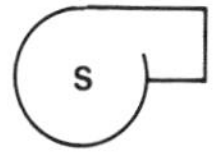

The mixed air is blown through filters and across heating/cooling coils, into humidifier and finally to supply diffusers.

The "air side" of the HVAC system offers the most opportunities to improve the quality of indoor air and is the easiest to address. Unfortunately, air-side maintenance is also the most labor intensive.

OUTDOOR AIR INTAKE

All air systems must have the capability to provide outdoor air to the building in order to replace air and dilute contaminants. The amount of outdoor air required varies with the type of facility and specific functions. For example, certain areas of a hospital, such as operating and delivery rooms, where contamination control is critical, have traditionally required continuous and total replacement of interior air with outdoor air. Research laboratories may also require total replacement.

The amount of outdoor air required to serve occupant needs is usually based on CO_2 content in the air. The new ASHRAE guidelines, 62-1989, aim to hold CO_2 below 1,000 ppm, or at 15 to 20 cfm/occupant in most facilities.

Outdoor air is drawn in, filtered, conditioned, passed through the space and exhausted from the building. The air intake location should allow good quality outdoor air to enter the facility and should be positioned so rain and snow do not enter the system. Just inside the air intake there should be a crude filter, a "bird screen," to prevent large objects from entering the system.

IAQ O&M OPPORTUNITIES:*

— check intake location to avoid:

- reentry of building exhaust,
- odors and contamination coming in from garbage dumpsters/compactors,

* Many of the IAQ O&M Opportunities cited in this chapter were derived from AQME's A Practical Maintenance Manual for Good Indoor Air Quality, an excellent manual for HVAC maintenance training.

- combustion contamination from indoor garages, loading docks, parking lots,
- exhaust from other buildings, and
- pollution from stagnant water;

— inspect bird screen to be sure it is free of obstruction at least twice a year (more frequently when intake is at ground level);

— be sure bird screen is intact;

— annually verify outdoor air locations and possible sources of contamination located near the intake;

— if there is any evidence of water penetration

- replace wet insulation with dry waterproof insulation, and
- install an indirect drain, change damper design if necessary;

— if occupants complain of stuffiness (or odors) check the quantity of outdoor air being admitted. To determine if it meets guidelines;

(1) count the number of occupants in an area and multiply by required cfm/person,

(2) obtain the recorded total volume of air being recirculated from a recent balancing report or by measuring with a Pitot tube,

(3) calculate the percentage of air required

$$\%\ Outdoor\ Air = \frac{No.\ of\ people * {}^{cfm}/_{person} * 100}{Total\ flow\ of\ system}$$

(4) with outdoor intakes in minimum position, measure the temperatures

— return temperature (RT) downstream from return fan

— mixing temperature (MT) at several points up stream from the supply fan; average the readings

— outdoor temperature (OT) in the vicinity of the intake,

It is best to do this when there is a significant temperature difference between outdoor air and return air; i.e., $90 < OT < 60$,

(5) calculate the outdoor air delivered by the system

$$\%OutdoorAir = \frac{RT-MT}{RT-OT} \times 100$$ and

(6) compare required outdoor air to delivered amount. If values differ, adjust. Wait approximately 10 minutes and repeat steps 4, 5, and 6.

MIXING PLENUM

The outdoor air and the recirculated air are mixed in the mixing plenum. Any dirt in this area can be carried into the ventilation system.

IAQ O&M OPPORTUNITIES:

— check for any dirt, dust, or moisture; clean the mixing plenum as needed;
— seal against any leakage; and
— verify that outdoor air is reaching plenum; clear air intake if warranted.

WATER AND AIR DISTRIBUTION SYSTEMS

Brief generic descriptions of water and air systems are offered to familiarize non-technical people with specific systems. These descriptions also provide a base of reference for the maintenance recommendations that follow. Those familiar with distribution systems may wish to turn directly to the discussion of O&M opportunities later in the chapter.

WATER SYSTEMS

Though many different air systems exist, there are basically only two types of water systems. While water system installation costs are usually higher, they are almost always more energy efficient and they avoid the contamination potential inherent in the long duct runs.

Unit Ventilator

In the unit ventilator, shown in Figure 9-2, the water moving through the coil is not controlled. However, the amount of air moving through the coil is controlled by the face and bypass damper. A thermostat and controller regulate the dampers. In this fashion, the air is cooled or heated as required.

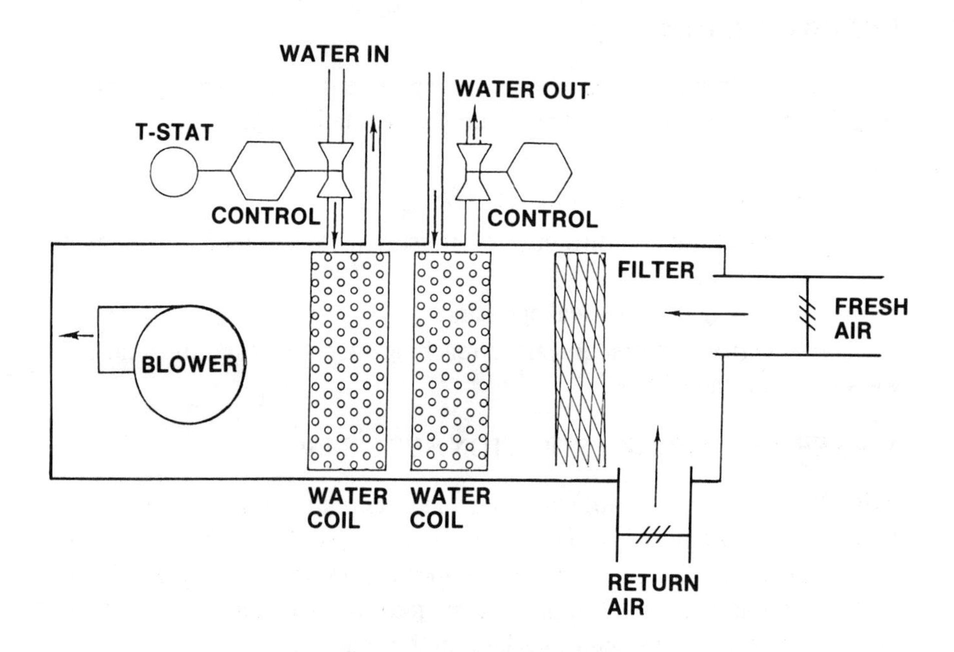

FIGURE 9-2. UNIT VENTILATOR; NO WATER VOLUME CONTROL

In addition to the face and bypass damper, the unit ventilator generally has two other dampers: the fresh air damper and the return air damper. These dampers may be set in a fixed position to allow a constant percentage of outdoor air to enter the room. They may also be controlled by a thermostat to allow for variable outdoor air percentages, depending upon the control scheme selected. Some unit ventilators have controls that allow for 100 percent outdoor air.

Fan Coil Unit

The second basic type of air moving unit is the fan coil unit as shown in Figure 9-3. The fan coil unit is the simplest of the various systems available for air movement and control. There are no dampers or controls inside the unit.

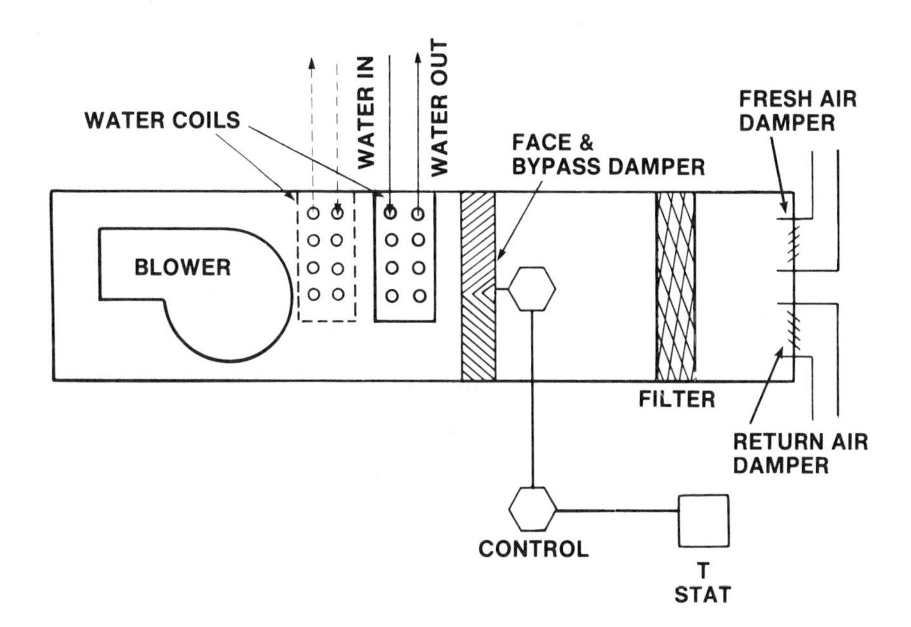

FIGURE 9-3. FAN COIL UNIT

Control of the fan coil unit is by means of a water valve which reacts directly to the thermostat setting. The control of water through the coil is the only built-in control for the temperature of the leaving air. Blower motor speed control is not commonly used. Fresh air is usually controlled by a damper in a fresh air duct.

The thermostat, not the air moving unit, should control the temperature of the air being delivered to the room through the unit ventilator or the fan coil. Adjustments to a thermostat should be made in very small increments with enough time allowed after each adjustment for the change to have an effect on the space temperature. Too great or too rapid a change in thermostat settings will frequently lead to over-correction of the space temperature.

Water Source Heat Pump

Heat pumps can pump heat out of a space or into a space (to or from the outdoors). Water source heat pumps have a water loop interjected between the pumps and the outdoors. Heat is pumped to or from this circulating water loop.

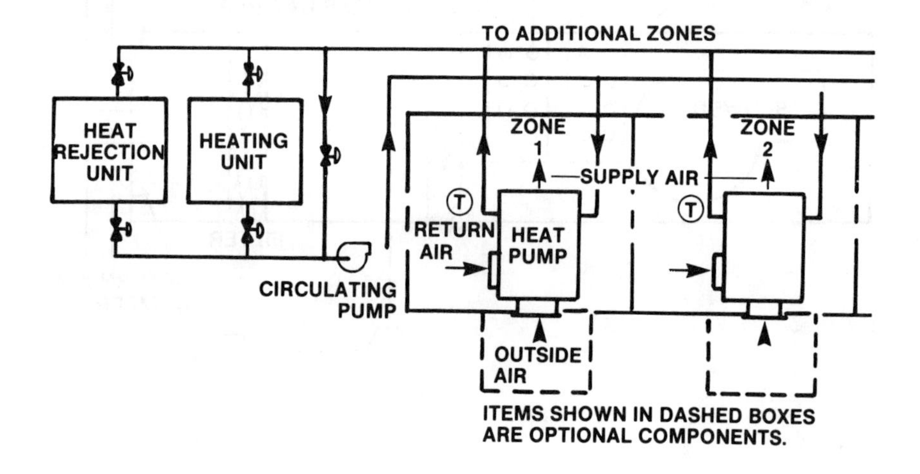

FIGURE 9-4. UNITARY HYDRONIC HEAT PUMP SYSTEM

AIR SYSTEMS

While air systems generally use chilled water from a central plant, they are sometimes outfitted with D-X or heat pump coils. There are many types of air systems. They fall in one of two categories: constant volume and variable air volume. The "constant volume" systems require a fixed amount of air to be distributed at all times. In the most common of these units, single and multizone, the duct work can be sized large enough to keep the air velocity (speed) low, resulting in a relatively low fan operating cost. Constant volume systems vary temperature to satisfy different load conditions.

Single Zone

In constant volume systems, air is moved to one or more locations via ducts. A single thermostat usually controls the amount of chilled or hot water going to the coil, or turns off/on electric heaters and turns off/on a refrigerant valve to the coil if D-X cooling is used. Normally, the fan always operates at the same flow rate. Auditoriums, dining halls, gymnasia and other large open spaces are usually equipped with ducted single zone systems.

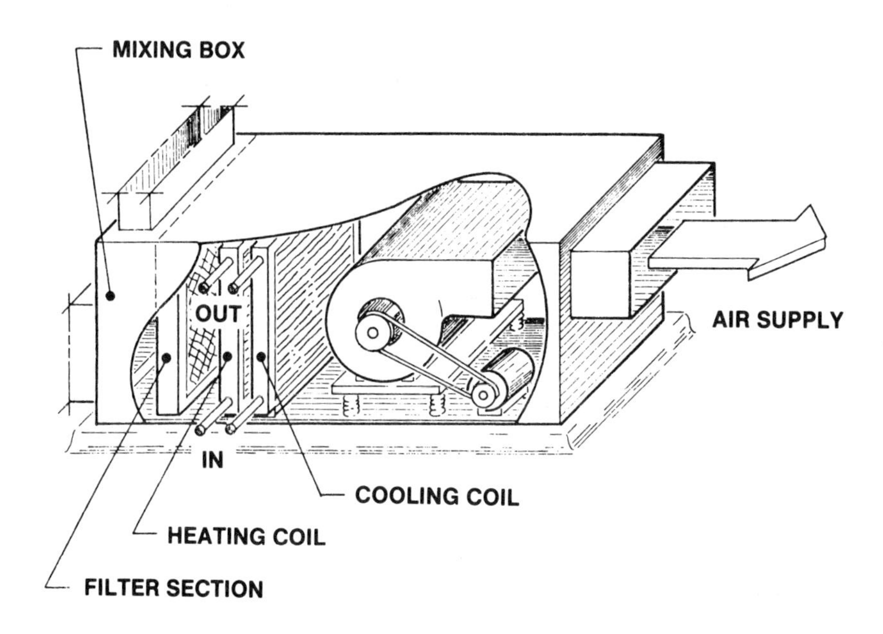

FIGURE 9-5. CONSTANT VOLUME, SINGLE ZONE

Single Zone with Terminal Reheat

An adaptation of the single zone system can be made with the addition of zone reheat coils and can offer some of the same advantages as a multizone system.

Multizone

The multizone air handler has one constant volume fan which supplies heated and cooled air to several individually controlled (thermostats) zones. The multizone system includes both the heating and cooling coils within the air handling unit (Figure 9-6). The multizone system controls different zone requirements by varying the amount of hot air and cold air to the zone. Cooled air is produced at constant temperature and then mixed with hot air to meet the room temperature required by the individual zone thermostat. Although this type of system can provide the consistent temperature desirable for indoor air quality, it is a costly process.

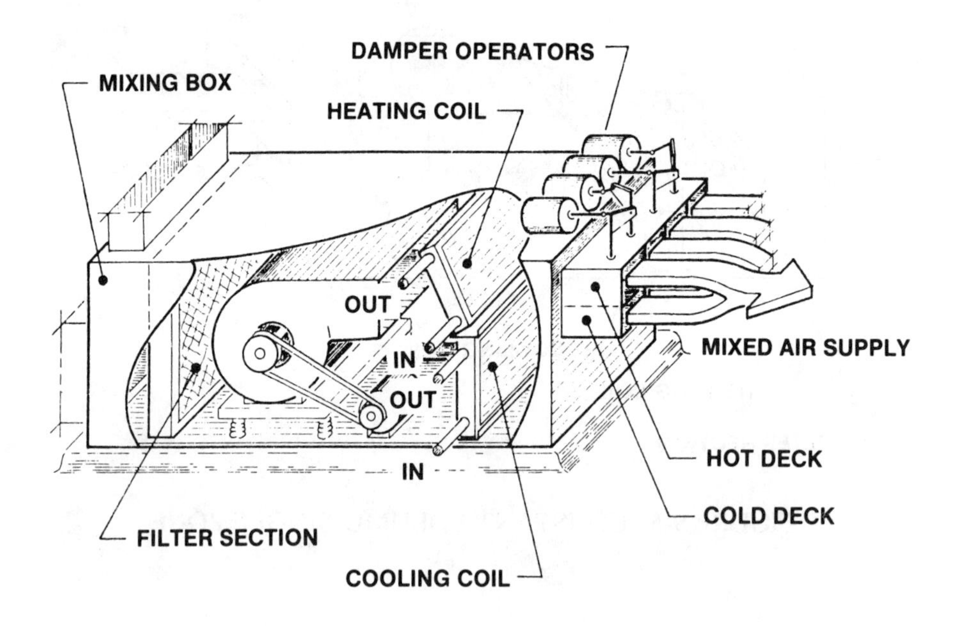

FIGURE 9-6. MULTIZONE SYSTEM

Dual Duct

The dual duct system is similar to the multizone system in that it has constant volume air supply and a hot and cold deck. In this system, however, air from the hot deck is ducted to each zone and air from the cold deck is ducted to the same zone. At each zone, in the plenum space, is a mixing box where the two air streams merge to provide tempered air according to demand from the thermostat.

Although this procedure provides excellent temperature control and a comfortable internal environment, it is hard on the utility bill. The duplication of duct work also presents some IAQ maintenance headaches, as a separate set of two duct runs is needed for each space.

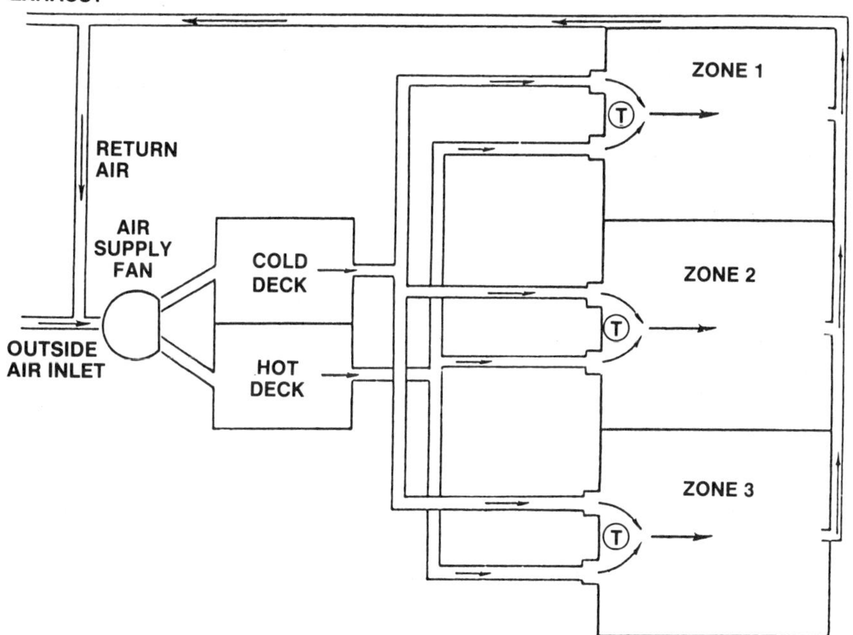

FIGURE 9-7. DUAL DUCT SYSTEM

Cooled air is mixed with heated air after both streams are ducted to each zone. Medium to high pressure ducts are used because the flow of air in each duct varies from 1 to 100 percent of the required air flow for each space.

All constant volume systems share the same generic drawback -- they move a fixed quantity of air at all times regardless of the actual building load requirements. In so doing, each of these systems (except the single zone unit) must operate on a "reheat" basis in which adding heat to the airstream is the only way to produce comfortable room conditions. As a result, constant volume systems must be designed large enough to handle the greatest load which may be required by all building zones simultaneously. The system's first cost is high since the equipment must be larger than is normally needed, and operating costs are likewise excessive.

The development of the "variable volume" air distribution system has provided designers a more efficient means of distributing comfort heating and cooling to building zones. This system is designed to meet all building load requirements while, at the same time, allowing energy savings at the fan and the heating/cooling plant.

Variable Air Volume

The variable air volume (VAV) system maintains temperature in a space by controlling the amount of heated or cooled air flowing to it. Conditioned air is delivered at medium to high pressure to the room through "VAV boxes" containing dampers which restrict the flow as the room thermostat dictates.

As a VAV system modulates from its cooling design level (maximum) to its heating design (minimum), total air circulation rate reduces. Since outdoor air is admitted as a constant percentage of the total air circulated, the outdoor air intake is reduced proportionately. Many VAV systems, however, have economizer cycle controls.

In a VAV system, the amount of outdoor air admitted at any one time depends upon:

(a) the total amount of air circulated (to which the percentage is applied); and

(b) the percentage of outdoor air brought in by the economizer cycle control.

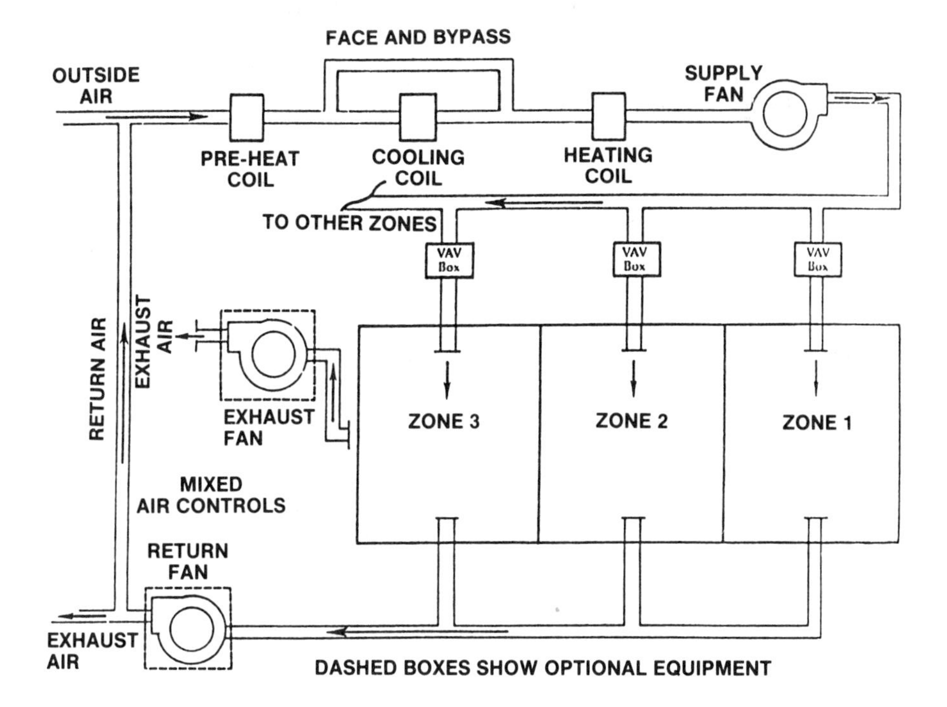

FIGURE 9-8. VARIABLE VOLUME FAN SYSTEM

If room temperature conditions are satisfied, the VAV could theoretically shut off the flow of air. However, health and safety reasons dictated, even during the energy crunch, that a minimum flow of air be maintained. VAV systems generally employ VAV boxes and medium pressure ducts. Since the late 1970s constant volume/low pressure duct systems have been retrofitted to have VAV characteristics.

- Ceiling by pass VAV -- Supply ducts to a zone are outfitted with damper boxes that "dump" supply air to the return air plenum as cooling becomes unnecessary. Thus, the air flow through the system is constant. The damper gradually varies the flow to the room or to the plenum as needed.

- Variable volume single zone -- instead of reheating cooled air in a single zone system, the air supply is varied by a variable speed drive or variable inlet valves in response to a pressure sensor located in the duct work.
- Side-pocket fan VAV system -- This augments the ceiling bypass or VAV single zone by providing a fan at the duct terminal. This fan draws air from the plenum and forces it through the register, along with the cooled air, into the zone. This maintains air circulation and air distribution patterns in the space.

All VAV systems as well as these hybrids govern air changes in response to temperature needs. Only indirectly do they respond to the number of occupants in a room; therefore, VAV systems are not as responsive to CO_2 levels or other contaminants.

In a facility without operable windows, a VAV system's effective ventilation rate to a number of spaces can vary radically. This is particularly true when indoor and outdoor ambient temperature are very close.

Since the VAV characteristic of throttling back outdoor air as a temperature response may result in unacceptable lower limits, some method to set a floor in outdoor air is needed. This requirement can be met by supplementing the VAV system with a constant-volume air supply in combination with the VAV system or independently. What would seem to be an obvious alternative, presetting for a minimum total air quantity could result in overcooling. Putting the supply air temperature at a higher value would increase total air flow, but at a cost of more energy for fan power.

Operations and maintenance opportunities related to key components, such as dampers, coils, fans, filters, are common to all the distribution systems and will be treated by component parts.

FANS

Indoor air quality is dependent on a fan to move air through the system. If fan blades are dirty or the belts are worn or stretched, fans will not work as effectively as they should and or may not run at their normal speed. When the static pressure and air flow are reduced, the air may not reach the last terminal box and areas at the "end of the line" do not receive sufficient air exchange.

IAQ O&M OPPORTUNITIES:

—check motor amps and volts with the fan running;
—listen for motor coast down noise;
—measure static pressure across the fan and check it for excessive vibration;
—determine if the last terminal box in the distribution system is receiving sufficient air flow;
—change fan rotation speed if warranted;
—clean blades with steam or a solvent annually;
—check the belts and pulley grooves for signs of wear or damage; and
—verify that belt tension is adequate; tighten or replace belts as needed.

Bearings

—inspect and lubricate fan and motor bearings;
—make sure greaseline connections and motor mounts are tight; and
—with fan in operation, make sure the shaft revolutions do not exceed the maximum.

DAMPERS

Dampers affect the amount of air intake, the amount recirculated and the amount exhausted. When dampers are not operating properly or are loose, air flow is not controlled and there may be unacceptable swings in temperature. In addition, the amount of "fresh" air may not be sufficient to assure indoor air quality. Conversely, improperly operating dampers may bring in too much air, which will not help IAQ and could require costly conditioning of unnecessary air.

IAQ O&M OPPORTUNITIES:

— check linkages, tightness and operation [Remember dampers are located throughout the system -- outdoor and return air systems, bypass ducts];

— make sure damper actuators are working properly; and

— repair, adjust and seal dampers as needed.

HEATING AND COOLING COILS

Heating coils warm the air to assure comfortable conditions for the occupants. If dirt obstructs the coil or the controls are not operating correctly, temperatures may be unacceptable. Control problems may stem from the design of the coil, a heating valve that is not functioning correctly, or a faulty or poorly adjusted control system.

Cooling coils cool and dehumidify the air supply. Since the coil dehumidifies the air, water gathers. In addition to heating coil problems, the cooling coil must be drained properly to avoid biological growth.

IAQ O&M OPPORTUNITIES:

— measure the heating coil discharge temperature regularly; if too high or too low, the control system needs to be checked;

— check if the water or steam is being closed off but air is still being heated. If this is occurring, the valve is not closing tightly;

— inspect and clean entire coils periodically; be sure coil fins are clean;

— make sure the off season cooling coils are "winterized;" and

— be sure condensate drains and drain pans are clean and operating properly.

HUMIDIFIERS

The humidifier supplies humidity to the air. The amount of humidity required is directly related to the quantity of outdoor air being brought into the facility, the degree to which the air must be warmed and the relative humidity (RH) of the outdoor air. Humidifier IAQ problems include stagnant water and water carryover into duct insulation. The prevalence of humidifier fever attests to the importance of cleaning and maintaining humidifiers.

IAQ O&M OPPORTUNITIES:

— check humidifier sprays, grids and pans for scale buildup and plugging;
— adjust the steam float valve;
— calibrate and verify the humidity sensor frequently;
— empty and clean pans when humidifier is not operating;
— make sure duct insulation is not wet; if wet, replace; and
— check for proper operation of humidifier controls, including high limit humidistat on controller.

FILTERS

In addition to the primary role filters serve in removing particles and fibers from the air, they reduce the microorganisms that attach themselves to these particles. Filters also prevent dust and dirt from accumulating on work surfaces, walls and equipment. This reduces custodial work and helps maintain equipment. It also relieves occupant concerns since cleanliness is usually associated with air quality. Filters also preserve the "health" of the entire ventilation system by preventing the accumulation of material downstreams in ducts, fans coils and other ventilation components. It is far more difficult and costly to fix downstream problems and to remove contaminates than to maintain an effective filtration system.

Air cleaner locations may vary with the purpose of the filter. If the contaminants are primarily from the occupied space, then it is logical to place the filter in the recirculated airstream. If filterable contaminants from outdoors are a concern, then filters can be placed in the supply air. Since outdoor air has been described as a natural reservoir for fungi and many occupants have reactions to pollen, the growing consensus is that outdoor air as well as recirculated air should be filtered.

Media filters and electrostatic air cleaners are typically found in building HVAC systems. Neither will function well without regular maintenance. As particles build up on the media, it will increase collection efficiency and screen out smaller particles. This efficiency, however, does not justify leaving media filters unattended. As particles build up, filters become clogged and the critical air flow rate decreases. Clogged filters increase resistance and reduce system efficiency. The dirtier the filter is when changed; the more bioaerosols will be disbursed into the air during the changing process.

Electrostatic air cleaners (EAC) will not operate properly unless absolutely clean. Build up on the charged plates interferes with the unit's effectiveness and, in fact, dirt on plates acts as pollutant sources.

Design considerations that effect filter maintenance include:

- Filter accessibility. Location of the unit, limited space, doors that are obstructed or held in place with screws all discourage filter changes.
- Filters that don't fit the opening. Filters that don't cover the opening cause a filter area by-pass into the open area where there is less resistance. Or, the gaps may be covered with other material, such as plywood. This can increase air velocity across filters, which, in turn, can increase static pressures and may decrease filter efficiency.

VAV systems present a special problem. Since the amount of filterable material is determined by volumetric flow and the filter's effectiveness, the reduced flow in a VAV system reduces contaminant removal capacity. Increasing the filter's effectiveness (information obtained from the manufacturer) can compensate for the reduced flow. It is also possible to insert additional filtration and recirculate the ventilation air to enhance filtration and compensate for reduced airflow.

Unless there is special filter expertise on staff, filters should be selected by specialists, who understand the problems and the options available. Guidelines can be found in ASHRAE 52-1976 and BOCA 62-73.

IAQ O&M OPPORTUNITIES:

— change filters regularly. Filter changes may be periodically scheduled or done in response to static pressure measurements;

— check and record static pressure loss at the filters regularly; when pressure reaches manufacturer/design level, the filters must be changed. (Be sure static pressure indicators are working properly.); and

— stop ventilation system to change filters wherever feasible.

If a fan continues to run while filters are being changed (in low pressure or low velocity systems where this is possible), material collected by the filters can escape and be distributed through the system. It is desirable to vacuum the filter area before the new filter is installed.

DUCTS

Air can travel literally miles and miles through duct work. Poorly maintained duct work can add pollution. (Dust near the diffusers may not be from upstream duct work but charged particles from the room.) Moisture in the duct work encourages microbiological growth, which can cause building related illness.

Fiberglass lined duct work can be cleaned if the problem is an accumulation of dust and spores. It cannot be effectively cleaned if mold growth on the fiberglass itself has occurred. Fiberglass lining should not be used in areas of high humidity or where water air washers are part of the system. Moisture and fiberglass-lined duct work almost always spell microbial contamination.

IAQ O&M OPPORTUNITIES:

To reduce the probability of mold growth in duct systems, SMACNA (Sheet Metal and Air-Conditioning Contractors National Association, Inc.) in its recent publication, Indoor Air Quality makes the following system design and maintenance recommendations:

— promptly detect and permanently repair all areas where water collection or leakage has occurred;

— maintain relative humidity at less than 60 percent in all occupied spaces and low velocity air plenums. During the summer, cooling coils should be run at a low enough temperature to properly dehumidify conditioned air;

— check for, correct, and prevent further accumulation of stagnant water under cooling deck coils of air handling units, through proper inclination and continuous drainage of drain pans;

— use only steam as the moisture source for humidifiers in the ventilation systems. Steam should not be contaminated with volatile amines (sometimes used as rust inhibitors);

— once contamination has occurred (through dust or dirt accumulation or moisture-related problems) downstream of heat exchange components (as in duct work or plenum), additional filtration downstream may be necessary before air is introduced into occupied areas; ...

— air handling units should be constructed so that equipment maintenance personnel have easy and direct access to both heat exchange components and drain pans for checking drainage and cleaning. Access panels or doors should be installed where needed; and

— non-porous surfaces where moisture collection has promoted microbial growth (e.g., drain pans, cooling coils) should be cleaned and disinfected with detergents, chlorine-generating slimicides (bleach), and/or proprietary biocides. Care should be taken to insure that these cleaners are removed before air handling units are reactivated.

Fiberglass insulation that has become wet in service should be removed and replaced to reduce the risk of mold growth, and to restore thermal and acoustical performance levels. This insulation may be very difficult to dry under normal operating conditions.

The cleaning of duct work, whether lined or not, is a difficult task. Flowing large quantities of air through the system is not generally effective for anything but large pieces of extraneous material in the system. This is due to the fact that boundary layers which form at the duct surface are at very low velocity. This makes the entrainment of dust particles in the air stream difficult, if not impossible.

TERMINAL BOXES/DIFFUSERS

The terminal box is the last step in air flow to the room. It reduces the pressures in the system to the room's atmospheric pressure. If the mechanism is jammed, air flow is restricted. Whether the terminal box uses a metal register, perforated grillwork or a valve, regular maintenance is required.

IAQ O&M OPPORTUNITIES:

- —verify operation periodically;
- —clean terminal boxes every three-four years on a regular PM program;
- —when heating is at the terminal box, check air flow at diffusers; and
- —if room is perennially not warm enough on a dual duct system, leakage may be occurring. If leak exceeds 10 percent, clean and replace joint gaskets. If off by much more than 10 percent, clean finned coils and make sure any heating coil valve is not leaking.

HEATING/COOLING PLANTS

The heating and cooling plants themselves present some concerns not addressed in the discussion of components.

Heating plants always have a level of incomplete combustion, which produces combustion contaminants. (See Chapter 4 and Appendix A.) The level of contaminants generated depend on the combustion characteristics of the burners and furnace, the type of fuel used, and the burner operating mode. Leaks allow these contaminants to escape and ultimately reach the occupants. Burner/fuel ratio is a key operating parameter that affects combustion products generated, the build up of deposits internally,and the quantity of smoke emitted from the stack. Boiler stack emissions may also enter the building through air intake.

IAQ O&M OPPORTUNITIES:

- —check heating plants in the fall and during the heating season;
- —seal cracks; and
- —check burner settings and make sure burner parts have not deteriorated to avoid high CO emissions.

Cooling plants that have elevated chilled water temperatures can contribute to indoor air problems. Adjusting chilled water temperatures upward as the load decreases can push relative humidity up to 65-80 percent in occupied spaces. This not only makes the occupants uncomfortable, but increases the likelihood of microbial growth. It has a particularly disastrous effect if there is already standing water or dirty filters.

FIRST, CHECK THE HVAC SYSTEM

All too often the "INDOOR AIR PROBLEM!" hysteria has a typical pattern. Early complaints are written off. Management only recognizes it has a problem after the staff is upset, absenteeism grows, productivity drops and employee emotions run high. Owners only recognize they have a problem after tenants move out and the space remains empty.

Faced with economic losses and damaged relationships plus the pall of threatened lawsuits, the pressure is on to find the culprit. In a quandary, owner/managers rush out to hire a consultant and the big hunt for the contaminating villain and its source begins. Unfortunately, the "cause" may never be identified.

If the HVAC system is responsible for nearly 60 percent of the building generated IAQ problems and if it has the potential to resolve up to 80 percent of the problems, the case seems clear. Why not clean house, the HVAC house, first? A basic generic HVAC engineering/maintenance approach can avoid some major IAQ headaches. Even if IAQ were not a consideration, this approach could pay for itself in operating efficiency and in the longer life of the equipment. A sound HVAC engineering/maintenance approach produces a "profit" in overcoming lost productivity ... and lost income.

Chapter 10

WHAT "THEY" SAY

While the field of indoor air quality is still considered by many to be rather embryonic, there are already many excellent sources of information and guidance. These sources offer owners and operating personnel a means of keeping abreast of this rapidly changing field.

Resources are available to the practitioner from international studies, reports and conference proceedings. Several agencies of the federal government are involved in indoor air quality and states are increasingly developing supportive materials. These agencies offer information, some provide rather definitive guidance and increasingly, they apply new laws and regulations.

Several associations have been very active in relation to IAQ concerns, particularly the American Conference of Governmental Industrial Hygienists and the American Society of Heating Refrigerating and Air-Conditioning Engineers. The newly formed Environmental Engineers and Managers Institute of the Association of Energy Engineers promises to be an excellent source of information.

The following discussion highlights some international, national and association resources where information of value to practitioners might be found. Sources of information pertaining to specific pollutants are cited at the end of the discussion of each contaminant in Appendix A.

INTERNATIONAL RESOURCES

Other areas of the world have been actively involved in IAQ matters for a number of years. Studies and reports from abroad have proven helpful in supporting many United States initiatives.

Sources of particular value to the practitioner are cited below along with recent publications of interest indicating the nature of the assistance available.

Commission of European Communities

The Joint Research Centre --Institute for the Environment

Directorate General for Science, Research and Development
Luxenbourg

Of particular interest, the series Indoor Air Quality and Its Impact on Man. Report No. 4, "Sick Building Syndrome A Practical Guide."

du Ministere de l Energie et des Resources du QuebecMontreal, Quebec

In particular, A Practical Maintenance Manual for Good Indoor Air Quality from AQME made available by the Ministere.

International Standards Organization (ISO)Geneva

Acoustics, Description and Measurement of Environmental Noise (1987)

Evaluation of Human Exposure to Whole-Body Vibration (1985)

International Thermal Comfort Standards (7730)

IRSST
(Institut de recherche en sante et en securite du travail du Quebec)
Montreal, Quebec

For office building owners and managers, Strategy for Studying Air Quality in Office Buildings.

Ontario Ministry of Labor
Government of Canada
Toronto, Ontario M&A 1T7
Canada

Report of the Inter-ministerial Committee on Indoor Air Quality

World Health Organization (WHO)

Health Aspects Related to Indoor Air Quality (1979)

Indoor Air Pollutants - Exposures and Health Effects (1982)

Indoor Air Quality - Research (1984)

Air Quality Guidelines - Organic Pollutants (1987)

FEDERAL RESOURCES

Periodically the Environmental Protection Agency (EPA) issues a list of major contaminants and the federal agencies working on various aspects of that contaminant. The list has been as long as 40 pages, attesting to the extensive IAQ activity at the federal level.

Aside from the research and reports, the agencies issue various guidelines, standards and regulations. Actions range from the Federal Home Loan Mortgage Corporation (Freddie Mac) standards for environmental hazards evaluation for one- to four-family mortgages to Occupational Safety and Health Administration (OSHA) guidelines for indoor air investigators.

Principle agencies involved in indoor air quality are the EPA, OSHA and the National Institute of Occupational Safety and Health (NIOSH). Additionally, the Consumer Products Safety Commission (CPSC) occasionally releases safety information; e.g., selection and maintenance of humidifiers. No single federal agency has jurisdiction over the quality of indoor air in non-industrial buildings.

In most IAQ areas, EPA serves as a lead agency and provides assistance on a range of concerns. NIOSH has taken the leadership in investigating problem buildings. OSHA is responsible for safeguarding workers' health in the workplace, and has focused most of its attention to date on industrial work environments. The U.S. Department of Housing and Urban Development has issued regulations specifying formaldehyde emission limitations in some building materials.

Federal agencies and their publications of particular interest to the practitioner are listed below.

Environmental Protection Agency (EPA)

A Citizen's Guide to Radon

Fact sheets, "Indoor Air Facts"

No. 3, Ventilation and Air Quality in Offices

No. 4, Sick Buildings

Directory of State Indoor Air Contacts

Exposure to Radon Daughters in Dwellings

The Inside Story: A Guide to Indoor Air Quality

National Primary and Secondary Ambient Air Quality Standards

Radon Reduction Methods - A Homeowner's Guide

National Institute of Occupational Safety and Health (NIOSH)

Indoor Air Quality Selected References

Manual of Analytical Methods

National Institute of Building Standards (NIBS)

Asbestos Model Specifications Guide

Occupational Safety and Health Administration

Air Contaminants - Permissable Exposure Limits

U.S. Dept. of Health and Human Services (HHS)

The Consequences of Involuntary Smoking

Electrical Power and Research Institute (EPRI)

Manual on Indoor Air Quality, EPRI.FM-3469

In addition, the Tennessee Valley Authority, the Bonneville Power Administration, the National Bureau of Standards, National Aeronautical and Space Administration, the General Services Administration and the U.S.

Department of Energy have been involved in various aspects of indoor air quality. These and other agencies may, from time to time, have helpful IAQ information for building owners, managers and operating personnel.

STATE RESOURCES

The National Governors' Association has periodically issued Indoor Air Pollution statements, reflecting the governors' concern about this growing problem. These statements have made recommendations as to actions the federal government and the states should take, and have fostered some activity in the states.

States vary widely in their involvement in indoor air quality. States active in the IAQ issue for several years include California, Maine, Minnesota, New Jersey, New York, Vermont and Wisconsin. Some states have been particularly active in certain areas. For example, the New York State Energy Office has been deeply involved in radon concerns for a number of years and is an excellent resource in this area. Many state and local governments have instituted antismoking legislation and/or ordinances for public buildings.

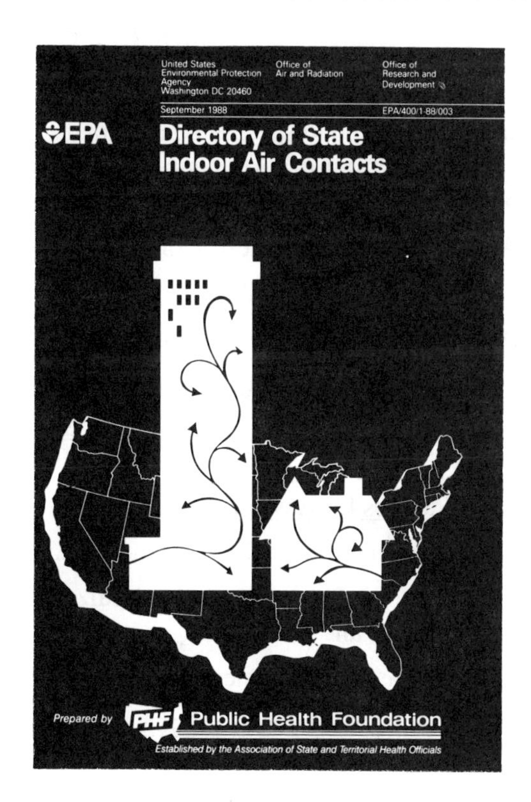

The practitioner can often find help on specific contaminants or general IAQ information at the state level; however, indoor air quality matters are not consistently handled by the same type of agency in each state. The lead agency may be public health, environment, energy or education. Furthermore, specific indoor air pollutants, such as asbestos, may be the responsibility of one agency while other IAQ matters are the purview of another agency, or agencies. There is, however, a Directory of State Indoor Air Contacts prepared by the Public Health Foundation and published by the EPA (1988) that can direct owners and operators to the appropriate source for obtaining information from their respective states. The directory is available from EPA, Air and Radiation Office, 401 M Street S.W., Washington, D.C. 20460.

In the absence of the directory, the department for health services is usually the best starting point. If that agency doesn't have the resources needed, someone there can usually refer you to a better source of information.

ASSOCIATIONS

A number of associations are active in the indoor air quality area. Aside from those devoted to specific contaminants, there are several that address the broad spectrum of IAQ concerns. Some of these organizations and publications of interest are listed below.

Air-Conditioning and Refrigeration Institute (ARI)

Briefing Paper on Indoor Air Quality

American Conference of Governmental Industrial Hygienists (ACGIH)

Threshold Limit Values and Biological Exposure Indices

A Manual of Recommended Practice for Industrial Ventilation (1989)

Guidelines for Assessing Bioaerosols (1990)

ASTM Subcommittee D 22.05, Indoor Air

Source of information on standardization of sampling and analyzing indoor air.

Air and Waste Management Association

Managing Asbestos in Schools, Public Commercial and Retail Buildings

Environmental Engineers and Managers Institute, Association of Energy Engineers

Strategic Planning for Energy and the Environment

American Society of Heating, Refrigerating and Air-Conditioning Engineers, Inc.

Standard 52-76 Method of Testing Air Cleaning Devices Used in General Ventilation for Removing Particulate Matter

Standard 90A-80 Energy Conservation in New Building Design

Standard 55-81 Thermal Environmental Conditions for Human Occupancy

Standard 62-89 Ventilation for Acceptable Indoor Air Quality

IAQ 86 Managing Indoor Air for Health and Energy Conservation

IAQ 87 Practical Control of Indoor Air Problems

IAQ 88 Engineering Solutions to Indoor Air Problems

IAQ 89 The Human Equation: Health and Comfort

Because of the potential sweeping impact ASHRAE's 62-89 standard is expected to have in the IAQ field, the history of this standard and some aspects of 62-89 warrant further discussion.

ASHRAE 62-1989, VENTILATION FOR ACCEPTABLE INDOOR AIR QUALITY

ASHRAE's historic leadership in comfort, health and matters related to indoor air quality cannot be captured in a few paragraphs. The Society's address is listed in the Resources and References for those who wish to learn more about its services and publications.

ASHRAE has had a dominant influence in establishing comfort and health standards in the United States for many years. The definition of sick buildings uses the 20 percent indicator from ASHRAE standard 55-81, which defined comfort in terms of conditions to satisfy 80 percent of the adult occupants.

ASHRAE's predecessor, The American Society of Heating and Ventilating Engineers, adopted its first ventilation standard in 1895 with 30 cubic feet per minute (cfm) as its minimum rate. In 1973, ASHRAE issued Standard 62-73, Standard for Natural and Mechanical Ventilation, which recommended ventilation rates for 140 applications with a minimum of 5 cfm/person and recommended rates for comfort in odor-free environments (typically 15 cfm/person). Modifications to 62-73 accommodated special rates for smoking, but found no reason to raise the minimum 5 cfm rate in its 62-81 revision, Ventilation for Acceptable Indoor Air Quality.

Soon after 62-81 was published, research results raised some doubts regarding the 5 cfm rate. One study found body odors persisted at 5 cfm, but were no longer perceived by people entering a room at 15 cfm. This 15 cfm rate lowered steady-state CO_2 levels to 1000 parts per million (ppm). U.S. Army studies revealed a 45 percent increase in respiratory infections among recruits in energy efficient buildings, which had only 5 percent outdoor air. Not 5 cfm; but 5 percent, or 1.8 cfm per occupant. The older barracks with 40 percent, which equated to 14.4 cfm per occupant, were found to have much lower infection rates.

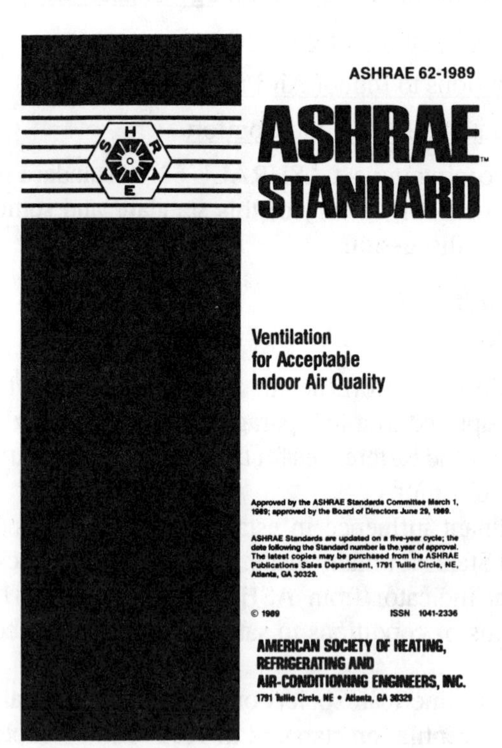

According to Janssen, chairman of the committee to revise 62-81, these two studies were sufficient to convince the committee that outdoor air flow should not be less than 15 cfm per person. The standard addresses most of the problems that cause poor indoor air quality and is a good reference regarding certain contaminants.

In an article summarizing the major changes in 62-89 compared to

62-81, Janssen noted the limitations in relying solely on cfm rates in the October 1989 ASHRAE Journal commenting;

> Dilution of tobacco smoke and other contaminants with less contaminated outdoor air is sometimes an imperfect control mechanism. It depends not only on the amount of dilution air, but on the degree of mixing achieved, convection currents, and perhaps other factors. Therefore, elimination of health risks through increased ventilation alone may not be possible.

ASHRAE 62-89 contains both prescriptive (Ventilation Rate Procedure) and performance (Air Quality Procedure) criteria. Compliance with both criteria need not be demonstrated. The implication is that compliance with one will satisfy the other.

VENTILATION RATES

While 15 cfm has now become the minimum under 62-89, many applications require more outdoor air. A major change from 62-81 is the removal of the distinction between cfm requirements in nonsmoking and smoking permitted areas. Table 10-1, taken from the new standard's Table 2, offers the cfm guidelines for a range of applications.

The standard does allow for some averaging to compensate for disparities in population density and varying room volumes. This averaging allows for a reduction in outside air to adjust for the ventilation requirements of the most demanding space.

Probably the greatest weakness in relying on a quantity of outside air for contaminant control is the uncertainty as to how effectively the air distribution serves the occupants. Since the air outlet and return air inlets are generally located at ceiling level in office buildings, the circulation may be short circuited and increased outside air will not have the desired effect. The standard's Table 2 assumes "well-mixed conditions". Section 6.1.3.3 speaks to ventilation effectiveness, but quantification procedures are not fully addressed in the standard.

62-89 does have a new provision, design documentation, which requires the designer to state the design assumptions regarding ventilation rates and air distribution. While this provision will help foster communications between designers and owners or operators, it raises further questions regarding responsibility and liability if the ventilation based on those assumptions proves inadequate. The standard states that the documentation "should be made available for operation of the system within a reasonable time after installation." Since the owner is ultimately responsible for the building's well-being, it may be prudent in today's litigious society to discuss these assumption with the designer prior to installation.

TABLE 10-1. (ASHRAE 62-1989)OUTDOOR AIR REQUIREMENTS FOR VENTILATION
2.1 COMMERCIAL FACILITIES (offices, stores, shops, hotels, sports facilities)

Application	Estimated Maximum Occupancy P/1000 ft^2 or 100 m^2	Outdoor Air Requirements cfm/person	cfm/ft^2	Comments
Dry Cleaners, Laundries				
Commercial laundry	10	25		Dry-cleaning processes may require more air.
Commercial dry cleaner	30	30		
Storage, pick up	30	35		
Coin-operated laundries	20	15		
Coin-operated dry cleaner	20	15		
Food and Beverage Service				
Dining rooms	70	20		
Cafeteria, fast food	100	20		
Bars, cocktail lounges	100	30		Supplementary smoke-removal equipment may be required.
Kitchens (cooking)	20	15		Makeup air for hood exhaust may require more ventilating air. The sum of the outdoor air and transfer air of acceptable quality from adjacent spaces shall be sufficient to provide an exhaust rate of not less than 1.5 cfm/ft2(7.5 L/s x m2).

TABLE 10-1. (ASHRAE 62-1989)OUTDOOR AIR REQUIREMENTS FOR VENTILATION - 2.1 COMMERCIAL FACILITIES continued

Application	Estimated Maximum Occupancy P/1000 ft^2 or 100 m^2	Outdoor Air Requirements		Comments
		cfm/ person	cfm/ft^2	
Garages, Repair, Service Stations				
Enclosed parking garage			1.50	Distribution among people must consider worker location and concentration of running engines; stands where engines are run must incorporate systems for positive engine exhaust withdrawal. Contaminant sensors may be used to control ventilation.
Auto repair rooms			1.50	
Hotels, Motels, Resorts, Dormitories			cfm/room	
Bedrooms			30	
Living rooms			30	
Baths			35	Installed capacity for intermittent use.
Lobbies	30	15		
Conference rooms	50	20		
Assembly rooms	120	15		
Dormitory sleeping areas	20	15		See also food and beverage services, merchandising, barber and beauty shops, garages.
Gambling casinos	120	30		Supplementary smoke-removal equipment may be required.

TABLE 10-1. (ASHRAE 62-1989)OUTDOOR AIR REQUIREMENTS FOR VENTILATION - 2.1 COMMERCIAL FACILITIES continued

Application	Estimated Maximum Occupancy P/1000 ft^2 or 100 m^2	Outdoor Air Requirements		Comments
		cfm/ person	cfm/ft^2	
Offices				
Office space	7	20		Some office equipment may require local exhaust.
Reception areas	60	15		
Telecommunication centers and data entry areas	60	20		
Conference rooms	50	20		Supplementary smoke-removal equipment may be required.
Public Spaces				
Corridors and utilities			0.05	
Public restrooms, cfm/wc or urinal		50		Mechanical exhaust with no recirculation is recommended. Normally supplied by transfer air, local mechanical exhaust; with no recirculation recommended.
Locker and dressing rooms			0.5	
Smoking lounge	70	60		
Elevators			1.00	Normally supplied by transfer air.
Retail Stores, Sales Floors, and Show Room Floors				
Basement and street	30		0.30	
Upper floors	20		0.20	

TABLE 10-1. (ASHRAE 62-1989)OUTDOOR AIR REQUIREMENTS FOR VENTILATION - 2.1 COMMERCIAL FACILITIES continued

Application	Estimated Maximum Occupancy P/1000 ft^2 or 100 m^2	Outdoor Air Requirements cfm/person	cfm/ft^2	Comments
Retail Stores, Sales Floors, and Show Room Floors cont.				
Storage rooms	15		0.15	
Dressing rooms			0.20	
Malls and arcades	20		0.20	
Shipping and receiving	10		0.15	
Warehouses	5		0.05	
Smoking lounge	70	60		Normally supplied by transfer air, local mechanical exhaust; exhaust with no recirculation recommended.
Specialty Shops				
Barber	25	15		
Beauty	25	25		
Reducing salons	20	15		
Florists	8	15		Ventilation to optimize plant growth may dictate requirements.
Clothiers, furniture			0.30	
Hardware, drugs, fabric	8	15		

TABLE 10-1. (ASHRAE 62-1989)OUTDOOR AIR REQUIREMENTS FOR VENTILATION - 2.1 COMMERCIAL FACILITIES continued

Application	Estimated Maximum Occupancy P/1000 ft^2 or 100 m^2	Outdoor Air Requirements		Comments
		cfm/ person	cfm/ft^2	
Specialty Shops cont.				
Supermarkets	8	15		
Pet shops			1.00	
Sports and Amusement				
Spectator areas	150	15		When internal combustion engines are operated for maintenance of playing surfaces, increased ventilation rates may be required.
Game rooms	70	25		
Ice arenas (playing areas)			0.50	
Swimming pools (pool and deck area)			0.50	Higher values may be required for humidity control.
Playing floors (gymnasium)	30	20		
Ballrooms and discos	100	25		
Bowling alleys (seating areas)	70	25		
Theaters				Special ventilation will be needed to eliminate special stage effects (e.g., dry ice vapors, mists, etc.)
Ticket booths	60	20		
Lobbies	150	20		

TABLE 10-1. (ASHRAE 62-1989) OUTDOOR AIR REQUIREMENTS FOR VENTILATION - 2.1 COMMERCIAL FACILITIES continued

Application	Estimated Maximum Occupancy P/1000 ft^2 or 100 m^2	Outdoor Air Requirements		Comments
		cfm/person	cfm/ft^2	
Theaters				
Auditorium	150	15		
Stages, studios	70	15		
Transportation				Ventilation within vehicles may require special considerations.
Waiting rooms	100	15		
Platforms	100	15		
Vehicles	150	15		
Workrooms				
Meat processing	10	15		Spaces maintained at low temperatures (-10°F to + 50°F, or -23°C to + 10°C) are not covered by these requirements unless the occupancy is continuous. Ventilation from adjoining spaces is permissible. When the occupancy is intermittent, infiltration will normally exceed the ventilation requirement. (See Ref 18).
Photo studios	10	15		
Darkrooms	10		0.50	
Pharmacy	20	15		

TABLE 10-1. (ASHRAE 62-1989) OUTDOOR AIR REQUIREMENTS FOR VENTILATION - 2.1 COMMERCIAL FACILITIES continued

Application	Estimated Maximum Occupancy P/1000 ft^2 or 100 m^2	Outdoor Air Requirements		Comments
		cfm/ person	cfm/ft^2	
Workrooms cont.				
Bank vaults	5	15		
Duplicating, printing			0.50	Installed equipment must incorporate positive exhaust and control (as required) of undesirable contaminants (toxic or otherwise).
2.2 INSTITUTIONAL FACILITIES				
Education				
Classroom	50	15		
Laboratories	30	20		Special contaminant control systems may be required for processes or functions including laboratory animal occupancy.
Training shop	30	20		
Music rooms	50	15		
Libraries	20	15		
Locker rooms			0.50	
Cooridors			0.10	
Auditoriums	150	15		
Smoking lounges	70	60		Normally supplied by transfer air. Local mechanical exhaust with no recirculation recommended.

TABLE 10-1. (ASHRAE 62-1989) OUTDOOR AIR REQUIREMENTS FOR VENTILATION - 2.2 INSTITUTIONAL FACILITIES continued

Application	Estimated Maximum Occupancy P/1000 ft^2 or 100 m^2	Outdoor Air Requirements		Comments
		cfm/person	cfm/ft^2	
Hospitals, Nursing and Convalescent Homes				
Patient rooms	10	25		Special requirements or codes and pressure relationships may determine minimum ventilation rates and filter efficiency. Procedures generating contaminants may require higher rates.
Medical procedure	20	15		
Operating rooms	20	30		
Recovery and ICU	20	15		
Autopsy rooms			0.50	Air shall not be recirculated into other spaces.
Physical Therapy	20	15		
Correctional Facilities				
Cells	20	20		
Dining halls	100	15		
Guard stations	40	15		

Reprinted by permission from ASHRAE Standard 62, copyright 1989, American Society of Heating, Refrigerating and Air-Conditioning Engineers, Atlanta, Georgia.

AIR QUALITY PROCEDURE

Recognizing the limits of outdoor air as an effective control mechanism, 62-89 kept the indoor air quality procedure from 62-81. This procedure does not specify the flow rate, but offers objective and subjective performance criteria, such as contaminant concentrations held below acceptable limits and odor acceptability. This procedure provides direct control of indoor air quality, but also poses a problem, as the standard offers little guidance on assessing, calculating or controlling pollutants to acceptable levels. Futhermore, acceptable limits have not been defined for many contaminants.

In some instances, guidance for indoor levels has been taken as 1/10 of the Threshold Limit Values (TLVs) as published by the ACGIH (1985). The standard does caution that 1/10 TLV may not provide an environment satisfactory to individuals who are extremely sensitive to an irritant.

Guidelines for some contaminants originating indoors are shown in Table 10-2.

TABLE 10-2. SELECTED AIR CONTAMINANTS ORIGINATING INDOORS

Contaminants	ppm	Time
Carbon dioxide	1,000	continuous
Chlordane	0.0003	continuous
Ozone	0.05	continuous
Radon	4 pCi/l[a]	1 yr. avg.

a This EPA recommendation applies specifically to residences and schools. ASHRAE recommends its use until guidelines for other facilities are published.

Recognizing that all outside air is not "fresh" air, 62-89 does provide for reduced outdoor air flow when the quality does not meet federal air quality standards. The National Ambient Air Quality Standard (NAAQS) for short-term concentration averaging is shown in Table 10-3. The standard applies the NAAQS contaminant levels to indoors for the same exposure times.

TABLE 10-3. NATIONAL AMBIENT AIR QUALITY STANDARDS

CONTAMINANT	CONCENTRATION AVERAGING					
	LONG TERM			SHORT TERM		
	$\mu g/m^3$	ppm	time	$\mu g/m^3$	ppm	time
Sulphur dioxide	80	0.03	1 yr	365	0.14	24 hrs[a]
Total particles[b]	50		1 yr	150	—	24 hrs[a]
Carbon monoxide				40,000	35	1 hrs[a]
Carbon monoxide				10,000	9	8 hrs[a]
Oxidants (ozone)				235[c]	0.12[c]	1 hrs[a]
Nitrogen dioxide	100	0.055	1 yr			
Lead	1.5	—	3 mos[d]			

a Not to be exceeded more than once per year.

b Arithmetic mean

c Standard is attained when expected number of days per calendar year with maximal hourly average concentrations above 0.12 ppm (235 $\mu g/m^3$) is equal to or less than 1, as determined by Appendix H to Subchapter C, 40 CFR 50

d Three-month period is a calendar quarter.

Source: U.S. EPA

When outside air does not meet NAAQS levels, the filtration system becomes a critical factor for make-up air and recirculated air. Should it become necessary to reduce outdoor air, more recirculated air will be required and the quality of filtration, contaminant sensing and control will become central to the process of maintaining healthy indoor air. In effect, poor quality outdoor air will force designers to use the indoor air quality procedure to comply with the standard.

Appendix C of the new standard discusses the range of limits for those pollutants where definite limits have not been set. For further background information, Appendix C offers Table C-1, Standards Applicable in the United States for Common Indoor Air Pollutants; Table C-2, Guidelines

Used in the United States for Common Indoor Air Pollutants; Table C-3, Summary of Canadian Exposure Guidelines for Residential Indoor Air Quality; and Table C-4, World Health Organization (WHO) Working Group Consensus of Concern About Indoor Air Pollutants at 1984 Levels of Knowledge. The tables offer an excellent set of references, but admittedly do not include all known contaminants that may be of concern. 62-89 leaves it to the user of the standard to decide if the contaminants not listed in its tables present a problem.

ACCEPTABLE AIR QUALITY

In an effort to comply with 62-89, owners and operators should note the reference to acceptable air quality as having "no known contaminants at harmful concentrations ... and with which a substantial majority (80 percent or more) of the people exposed do not express dissatisfaction." This level of satisfaction and acceptability (80 percent) clearly means that conditions do not have to be unanimously or universally applicable. In other words, to comply with the standard the space does not have to be conditioned to meet the unique needs of the hypersensitive -- or the perennial complainer.

One alternative suggested in 62-89 is the use of an odor panel. Appendix C describes how one type of odor panel might serve this purpose. The problem with this approach is the rapidity with which people become inured to an odor even when the source continues to present a health problem. A judgment of acceptability must be rendered in 15 seconds to be valid. Research has actually shown that odor and irritation are mutually inhibitory; therefore, the existence of one can lessen the perception of the other. An odorant, therefore, could be removed and the irritation could still increase.

The most manageable way to use the indoor air quality procedure would appear to be CO_2 monitoring. Since the 15 cfm rate is predicated on keeping CO_2 under 1000 ppm, procedures to keep CO_2 below 1000 ppm (and ventilation at an acceptable level) can be gauged using detector tubes. The indoor air quality procedure monitored for CO_2 levels can assure acceptable ventilation in a more energy efficient manner, using ventilation as needed; not as some blind volume approach. By the same token, if CO_2 can be held below 825 ppm in offices, the 20 cfm guidelines can be satisfied without dollars to condition unnecessary outdoor air. 62-89 does warn, however,

"In the event CO_2 is controlled by any method other than dilution, the effects of the possible elevation of other contaminants must be considered."

An underlying problem with cfm/occupant and CO_2 as an indirect measure is that the cfm guidelines rest on per occupant conditions. The CO_2 and odor surrogates are rooted in historical concerns regarding pollution by occupants. For hundreds of years, the focus has been on body odors, tobacco smoke and "stuffy" crowded rooms. We are still fighting the same old battle, but the rules of war have changed.

Many of our pollutant enemies on today's battlefield are not occupant-related. Furniture will off-gas at much the same rate if there are 2 or 20 people in a room. Bioaerosols can amplify in moldy carpets without help from people. Operating copiers and VDTs will pollute at the same rate with a few or a crowd. To the extent that 62-89 has not been able to address these concerns, it does not set a health standard for indoor air quality. Our needs have moved beyond the body/tobacco stage; some procedure for sensing and controlling contaminants not related to occupancy in an energy efficient manner must be found.

Until we know enough about the health effects and control strategies for all significant contaminants, compliance with 62-89 may be the only alternative. At this point, it would seem to be the legally prudent course to take. But it will not be easy or cheap, as Burroughs observed, "Designers and owners can, however, expect [cost] increases in all phases of construction due to 62-89 design; equipment; T&B; maintenance; commissioning and energy."

The ultimate irony may be that we will improve our indoor air quality at the expense of our outdoor climate. Owners and managers are apt to comply with ASHRAE 62-89 to provide some assurance of indoor air quality and to be legally prudent -- whether a building seems "sick" or not. Yet, the highest estimates of potentially sick buildings reaches only 30 percent; and, of those, NIOSH's experience would suggest less than half require more outside air. A three to four-fold increase in outside ventilation requirements is an expensive cure for only 15 percent of our building stock. The highest price of all, however, may be the significant increase in CO_2 emissions, accelerating the global climate change so many already fear.

SELECTED RESOURCES AND REFERENCES

RESOURCES

American Conference of Governmental Industrial Hygienists (ACGIH)
6500 Goenway Avenue, Building D-7,
Cincinnati, Ohio 45211

American Industrial Hygiene Association
475 Wolf Leges Parkway
Akron, Ohio 44311

American Society of Heating, Refrigerating and Air-Conditioning Engineers, Inc. (ASHRAE)
1791 Tullie Circle, N.E.
Atlanta, Georgia 30329

AQME
1259 rue Berri
bureau 510
Montreal, Quebec H2L 4C7
Canada

Association of Energy Engineers
4025 Pleasantdale Road, Suite 420
Atlanta, Georgia 30340

Commission of the European Communities
Joint Research Centre
Institute for the Environment
Ispra, Italy

Consumer Product Safety Commission
5401 Westbard Avenue
Bethesda, MD 20816

Environmental Engineers and Management Institute (AEE)
4025 Pleasantdale Road, Suite 420
Atlanta, Georgia 30340

Girman, John
Indoor Air Quality Program
Department of Health Services
State of California
2151 Berkeley Way
Berkeley, CA 94704

IRSST
505 boulevard de Maisonneuve Ouest
Montreal, Quebec H3A 3C2
Canada

National Environmental Balancing Bureau
8224 Old Courthouse Road
Vienna, Virginia 22180

National Institute of Occupational Safety and Health
Hazard Evaluations and Technical Assistance Branch
4676 Columbia Parkway
Cincinnati, OH 45226

Occupational Safety and Health Administration
Department of Labor
200 Constitution Ave, N.W.
Washington, D.C. 20001

Office of Smoking and Health
U.S. Public Health Service
5600 Fishers Lane, Rm 1-10
Rockville, Maryland 20857

U.S. Environmental Protection Agency
401 M Street S.W.
Washington, D.C. 20460

NEWSLETTERS

Indoor Air Quality Update
Cutter Information
1100 Massachusetts Ave.
Arlington, MA 02174

Indoor Pollution Law Report
Leader Publications
111 Eighth Ave.
New York, NY 10011

Indoor Pollution News
Buraff Publications Inc.
1350 Connecticut Ave., N.W.
Suite 1000
Washington, D.C. 20036

Pollution Equipment News
Rimbach Publications
8650 Babcock Blvd.
Pittsburgh, PA 15237

REFERENCES

ACGIH. ACGIH Threshold Limit Values and Biological Exposure Indices for 1986-87, American Conference of Governmental Industrial Hygienists, Cincinnati (1987).

Amman, H.M. and M.A. Berry, N.E. Childs, D.T. Mage. "Health Effects Associated with Indoor Air Pollutants." Managing Indoor Air for Health and Energy Conservation, (1986), 53-70.

Anne Arundel County Public Schools, Indoor Air Quality Management Program. Annapolis: AACPS, 1989.

ASHRAE 52-1968 (RA76), Methods of Testing Air-Cleaning Devices Used in General Ventilation for Removing Particulate Matter.

ASHRAE 55-1981, Thermal Environmental Conditions for Human Occupancy.

ASHRAE 62-1989, Ventilation for Acceptable Indoor Air Quality.

ASHRAE 90.1-1989, Energy Efficient Design of New Buildings Except New Low-Rise Residential Buildings.

Bahnfleth, D.R. and F.A. Govan. "Effect of Building Airflow on Reentry and Indoor Air Quality." Practical Control of Indoor Air Problems (1987), 185-193.

Ball, James and Myron Webb. Interior Landscape Plants for Indoor Air Pollution Abatement. Washington, D.C.: National Aeronautics and Space Administration, 1989.

Berglund, L.G. and W.S. Cain. "Perceived Air Quality and the Thermal Environment." The Human Equation: Health and Comfort, (1989), 93-99.

Berlin, Gary. "Indoor Air Quality: Control Humidty to Increase Productivity." Contracting Business (June 1989).

Bernard, J.M. "Building-Associated Illnesses in an Office Environment.: Managing Indoor Air for Health and Energy Conservation, (1986), 44-52.

Burge, H.A. "Environmental Allergy: Definitions, Causes, Control." Engineering Solutions to Indoor Air Problems, (1988), 3-7.

Burroughs, H.E. "Forward" Engineering Solutions to Indoor Air Problems (1988), vii-ix.

Burroughs, H.E. "Indoor Air Quality: The Strategic Challenge of the Nineties." Paper presented 12th World Energy Engineering Congress, Association of Energy Engineers, (1989).

Cain, William S. and Brian P. Leaderer. "Ventilation Requirements in Occupied Spaces During Smoking and Nonsmoking Occupany." Environment International 8 (1982), 505-514.

Committee on Passive Smoking, National Research Council, and National Academy of Sciences. Environmental Tobacco Smoke Measuring Exposures and Assessing Health Effects. Washington, D.C. 1986.

Cyfracki, Ludwik. "Benefits Exceed Costs for Quality Heating, Ventilating and Air Conditioning Systems (HVAC)." Paper presented Forum Energie (1988).

Duffy, Gordon. "Indoor Air Quality Takes a More Practical Bent." Engineered Systems, (Nov/Dec 1988), 43-48.

Fleming, W.S. "Indoor Air Quality, Infiltration, and Ventilation in Residential Buildings." Managing Indoor Air for Health and Energy Conservation, (1986), 192-207.

Fuestel, H.E. and M.P. Modera, A.H. Rosenfeld. "Ventilation Strategies for Different Climates." Managing Indoor Air for Health and Energy Conservation, (1986), 342-363.

Gardner, Thomas F. "IAQ: Legal Trouble for the 90s." Air Conditioning, Heating and Refrigeration News, (March 5 1990), 50-51.

Green, George H. "The Effect of Indoor Relative Humidity on Absenteeism and Colds in Schools." ASHRAE Journal, (Jan 1975), 57-62.

Godish, Thad. Indoor Air Pollution Control. Chelsea, MI: Lewish publishers, 1990. [Focused on residential and small business prevention and control strategies.]

Goldhaber, Marilyn K., Michael R. Plen and Robert A. Hiatt. "The Risk of Miscarriage and Birth Defects Among Women Who Use Visual Display Terminals During Pregnancy." American Journal of Industrial Medicine, 13 (1988), 695-706.

Groen, Doug. "Sick Building Syndrome -- Causes, Effects and Cures." ECON, (June 1988), 50-52.

Hall, Stephan K. and Sue A. Lavite. "Indoor Air Quality in Commercial Buildings." Pollution Engineering, (June 1988), 54-59.

Hodgson, M.J. and K. Kreiss. "Building-Associated Diseases: An Update." Managing Indoor Air for Health and Energy Conservation, (1986), 1-13.

Janczewski, Jolanda and Jon M. Yarek. "Indoor Air Quality: Should You Be Concerned?" Facilities Manager (Summer 1989), 22-29.

Janssen, John E. "Ventilation for Acceptable Indoor Air Quality." ASHRAE Journal, (October 1989), 40-48.

Lane, C.A. and J.E. Woods, T.A. Bosman. "Indoor Air Quality Diagnostic Procedures for Sick and Healthy Buildings." The Human Equation: Health and Comfort, (1989), 237-240.

Levin, H. "IAQ-Based HVAC Design Criteria." Engineering Solutions to Indoor Air Problems, (1988), 61-68.

Lubart, J. "The Common Cold abnd Humdity Imbalance." NYS Journal of Medicine, (1962), 816-819.

McNall, P.E. "Control of Indoor Air Quality by Means of HVAC Systems." Managing Indoor Air for Health and Energy Conservation, (1986), 541-547.

Mikulina, Thomas W. "Increasing the Priority of Comfortability." The NESA Newsletter, Viewpoint (Sept 1989), 4.

Morey, P.R. "Microorganisms in Buildings and HVAC Systems: A Summary of 21 Environmental Studies," Engineering Solutions to Indoor Air Problems, (1988), 10-21.

Penz, Alton. Vice President for Research, Building Owners Management Association (BOMA). In conversation, 1990.

Persily, A.K. "Control Technology and IAQ Problems." Engineering Solutions to Indoor Air Problems, (1988), 51-59.

Proetz, A.W. Annals of Otology, Rhinology and Larynology, 65 (1956), 376-384.

Rask, Dean R. and Charles A. Lane. "Resolution of Sick Building Syndrome: Part II, Maintenance." The Human Equation: Health and Comfort. (1989), 173-177.

Robertson, Gray. "Ventilation, Health and Energy Conservation -- A Workable Compromise." Strategic Planning and Energy Management, (Spring 1989), 65-78.

Skov, Peder and Ole Valbjorn. "The Sick Building Syndrome in the Office Environment: The Danish Town Hall Study." Environmental International, 13 (1987), 339-349.

Spengler, John D. and Haluk Ozkaynak, John F. McCarthy, Henry Lee. Summary of Symposium of Health Aspects of Exposure to Asbestos in Buildings. Boston: Harvard School of Public Health, 1989.

Surgeon General's Report. The Health Consequences of Involuntary Smoking, Washington, D.C.: U.S. Public Health Service. 1986.

U.S. EPA. National Primary and Secondary Ambient Air Quality Standards. CFR, Title 40 Part 50 as amended July 1, 1987.

Vaculik, F. "Air Quality Control in Office Buildings by CO_2 Method." Practical Control of Indoor Air Problems, (1987), 244-249.

Wellford, B.W. "Mitigation of Indoor Radon Using Balanced Mechanical Ventilation Systems." Managing Indoor Air for Health and Energy Conservation, (1986), 602.

GLOSSARY OF TERMS

To maintain consistency, the following terms in common with ASHRAE 62-1989 definitions use the terminology which appears in the standard.

absorption: the process of one substance entering into the inner structure of another.

acceptable indoor air quality: air in which there are no known contaminants at harmful concentrations as determined by cognizant authorities and with which a substantial majority (80% or more) of the people exposed do not express dissatisfaction.

adsorption: the adhesion of a thin film of liquid or gases to the surface of a solid substance.

air-cleaning system: a device or combination of devices applied to reduce the concentration of airborne contaminants, such as microorganisms, dusts, fumes, respirable particles, other particulate matter, gases, and/or vapors in air.

air-conditioning: the process of treating air to meet the requirements of a conditioned space by controlling its temperature, humidity, cleanliness, and distribution.

air, ambient: the air surrounding an object.

air, exhaust: air removed from a space and not reused therein.

air, makeup: outdoor air supplied to replace exhaust air and exfiltration.

air, outdoor: air taken from the external atmosphere and, therefore, not previously circulated through the system.

air, recirculated: air removed from the conditioned space and used for ventilation, heating, cooling, humidification, or dehumidification.

air, return: air removed from a space to be then recirculated or exhausted.

air, supply: that air delivered to the conditioned space and used for ventilation, heating, cooling, humidification, or dehumidification.

air, transfer: the movement of indoor air from one space to another.

air, ventilation: that portion of supply air that is outdoor air plus any recirculated air that has been treated for the purpose of maintaining acceptable indoor air quality.

allergen: a substance that induces allergic reaction.

arthralgia: neuralgic pain in one or more joints.

chemisorb: to take up and hold, usually irreversibly, by chemical forces.

concentration: the quantity of one constituent dispersed in a defined amount of another.

conditioned space: that part of a building that is heated or cooled, or both, for the comfort of occupants.

contaminant: an unwanted airborne constituent that may reduce acceptability of the air.

dilution ventilation: dilution of contaminated air with uncontaminated air in a general area, room, or building for the purpose of health hazard or nuisance control. (ACGIH, 1984)

dose: the amount of exposure undergone at one time.

dust: an air suspension of particles (aerosol) of any solid material, usually with particle size less than 100 micrometers.

dyspnea: shortness of breath; difficult or labored respiration.

energy recovery ventilation system: a device or combination of devices applied to provide the outdoor air for ventilation in which energy is transferred between the intake and exhaust airstreams.

epidemiology: a branch of medicine that investigates the causes and control of epidemics; all the elements contributing to the occurrence or nonoccurrence of a disease in a population; ecology of a disease.

ergonomics: the study of people adjusting to their work environment; the science of adapting working conditions to the worker.

etiology: the science of causes or origins; the causes of a specific diseases.

exfiltration: air leakage outward through cracks and interstices and through ceilings, floors, and walls of a space or building.

fatigue: physical or mental exhaustion; weariness; tiredness.

fumes: airborne particles, usually less than 1 micrometer in size, formed by condensation of vapors, sublimation, distillation, calcination, or chemical reaction.

gas: a state of matter in which substances exist in the form of nonaggregated molecules, and which, within acceptable limits of accuracy, satisfies the ideal gas laws; usually a highly superheated vapor.

hazard: risk, peril, jeopardy to which an individual is subjected.

infiltration: air leakage inward through cracks and interstices and through ceilings, floors, and walls of a space or building.

inhalable: particles small enough to be inhaled, but large enough so they are not quickly exhaled.

lethargy: a condition of abnormal drowsiness or torpor; a great lack of energy; apathy.

malaise: a vague feeling of discomfort or uneasiness.

micron (μ): a unit of linear measure equal to one millionth of a meter, or one thousandth of a millimeter.

microorganism: a microscopic organism, especially a bacterium, fungus, or a protozoan.

myalgia: pain in one or more muscles.

natural ventilation: the movement of outdoor air into a space through intentionally provided openings, such as windows and doors, or through nonpowered ventilators or by infiltration.

occupied zone: the region within an occupied space between planes 3 and 72 in. (75 and 1800mm) above the floor and more than 2 ft (600mm) from the walls or fixed air-conditioning equipment.

odor: a quality of gases, liquids, or particles that stimulates the olfactory organ.

oxidation: a reaction in which oxygen combines with another substance.

particulate matter: a state of matter in which solid or liquid substances exist in the form of aggregated molecules or particles. Airborne particulate matter is typically in the size range of 0.01 to 100 micrometers.

plug flow: a flow regime where the flow is predominately in one direction and contaminants are swept along with the flow.

polyuria: excessive secretion of urine.

presumptive: giving reasonable cause for belief; presumed.

smoke: the airborne solid and liquid particles and gases that evolve when a material undergoes pyrolysis or combustion. Note: chemical smoke is excluded from this definition.

total suspended particulate matter: the mass of particles suspended in a unit of volume of air when collected by a high-volume air sampler.

respirable particles: respirable particles are those that penetrate into and are deposited in the nonciliated portion of the lung. Particles greater than 10 micrometers aerodynamic diameter are not respirable.

toxic: of, affected by, or caused by a toxin; to cause a poisonous reaction.

vapor: a substance in gas form, particularly one near equilibrium with its condensed phase, which does not obey the ideal gas laws; in general, any gas below its critical temperature.

ventilation: the process of supplying and removing air by natural or mechanical means to and from any space. Such air may or may not be conditioned.

GLOSSARY OF ACRONYMS AND ABBREVIATIONS

ACGIH	American Conference of Governmental Industrial Hygienists
ACH	air changes per hour
AEE	Association of Energy Engineers
ANSI	American National Standard Institute
ASHRAE	American Society of Heating, Refrigerating and Air-Conditioning Engineers
BOCA	Building Officials and Code Administrators
BOMA	Building Owners Management Association
cfm	cubic feet per minute
CPSC	Consumer Product Safety Commission
DOE	United States Department of Energy
EEMI	Environmental Engineers and Managers Institute (AEE
EPA	United States Environmental Protection Agency
ETS	environmental tobacco smoke
HEPA	high efficiency particulate air filters
HVAC	heating, ventilating and air-conditioning (system)

IAQ	indoor air quality
IAQU	Indoor Air Quality Update, Newsletter of Cutter Information Corp., Arlington, MA
NAAQS	National Ambient Air Quality Standards
NESHAP	National Emissions Standards for Hazardous Air Pollutants
NIOSH	National Institute of Occupational Safety & Health, U.S. Department of Health & Human Services
mg/m^3	milligrams per cubic meter
OSHA	Occupational Safety & Health Administration
PCBs	polychlorinated biphenyls
pCi/l	picocuries per liter; a measure of radon concentration
PELs	permissable exposure limits
Picocuries	(pico = one trillionth) a unit of radioactive nuclide in which 3.7 X 1010 disintegrations occur per second
ppb	parts per billion
ppm	parts per million
RH	relative humidity; ratio of amount of water vapor in the air relative to the greatest amount possible at the same temperature
RSP	respirable suspended particles
TLVs	threshold limit values
TSP	total suspended particulate concentration
ug/m^3	micrograms per cubic meter
um	micrometer

VAV	variable air volume
VOC	volatile organic compound
WEEL	workplace environmental exposure limit
WHO	World Health Organization
WL	working level; a unit of radon exposure, a weighted average of the concentration of radon daughters

APPENDIX A
CONTAMINANTS

ASBESTOS

DESCRIPTION

Asbestos is a term used to refer to a number of inorganic minerals that have specific properties in common. The serpentine mineral, chrysotile, is the most commonly used asbestos and represents about 95 percent of the asbestos used in buildings in the United States. The second largest asbestos group is amphiboles, which includes amosite and crocidolite. The fiber structure and the associated health risks are very different by type.

Once in place, asbestos does not degenerate spontaneously. Fiber bundles are not inclined to be disrupted without some mechanical external force.

SOURCES

Man has used asbestos for thousands of years. Until the 1970s, it was the material of choice for thermal and acoustical insulation as well as fire-proofing. It can be found in thousands of commercial products including reinforced cement, thermal insulation, floor tiles, gaskets, brake linings, and heat-resistant textiles.

In the building structure, asbestos containing material (ACM) is most frequently found in boiler insulation, pipe insulation, sprayed-on fire proofing, breaching insulation, floor and ceiling tiles. The EPA estimates that 20 percent, or 733,000, buildings in the country contain friable asbestos -- and this figure excludes schools and residential buildings with less than ten units.

SYMPTOMS AND HEALTH EFFECTS

There are no immediate discernable symptoms of asbestos exposure. EPA and medical specialists have estimated that 1,000 to 7,000 already exposed people will die of asbestos-related diseases over the next 30 years.

All current information on the health effects resulting from asbestos inhalation come from studies of occupational settings with high exposure levels. Most data are based on the amphiboles group, not common in U.S. construction. Three forms of disease have been associated with the inhalation of asbestos fibers: (1) asbestos-fibrosis or scarring of the lungs; (2) mesothelioma - a malignancy of the linings of the lung and abdomen; and (3) lung cancer.

Asbestosis - deaths have only been observed among individuals that have been occupationally exposed to high levels of asbestos. There does not appear to be any evidence that asbestosis should be a concern as a result of environmental exposures today.

Mesothelioma and lung cancer - conservative assessments by various researchers, including Liddel, Hughes and Peto, place the associated lifetime risk of death at 1 per 100,000 for 10 years of occupancy in buildings with 0.001 regulatory fibers/ml of mixed fiber types. Recent data suggest average level of asbestos in schools and other buildings with ACM is generally well below 0.001 fibers/ml.

The primary occupant health risks would appear to be to operations and maintenance people who, in the course of their duties, disturb ACM. The greatest health risks are to those who remove the material.

Mounting evidence that asbestos is not as great a problem as previously surmised prompted Congress to order the EPA in 1989 to have the Health Effects Institute reassess asbestos-related health risks.

LAW/REGULATIONS/STANDARDS

Schools - 1982 Asbestos in Schools Rule and 1986 Asbestos Hazard Emergency Response Act (AHERA) applies to schools. 40 CFR Part 763, Part III "Asbestos-Containing Material in Schools," April 30, 1987

Guidelines for removal, OSHA Title 29, CFR 1910.1001 and 1926.58, U.S. Department of Labor

EPA, Industrial and Respiratory Division, bans on certain applications and guidelines

National Emissions Standards for Hazardous Air Pollutants (NESHAPS) 40 CFR, Subpart M, Section 61.47

State codes

Local ordinances

ACCEPTABLE LEVELS

Laws/regulations/standards noted in 3 above (and any subsequent revisions) establish levels and conditions of use.

After a careful review of the existing studies and their findings, the Harvard Asbestos Symposium concluded, "At present there is insufficient information to correlate the relationship between potential exposure and actual indoor concentrations of airborne asbestos fibers."

MONITORING AND MEASUREMENT

EPA regulatory programs rely on visual inspection methods. Many asbestos specialists feel the training being offered in the field is inadequate to consistently and accurately assess ACM conditions and potential disturbance as needed to fulfill regulatory requirements. There is a lack of correlation between this visual approach and fiber concentrations in spaces with damaged ACM. Since visual inspections do not quantify exposure. This method does not offer reliable risk assessment.

School measurement procedures must comply with AHERA's regulations.

To estimate asbestos exposure and associated risks, the 1989 Harvard Symposium on Health Aspects of Exposure in Buildings concluded that it is necessary to characterize airborne fiber concentrations - ideally by length and diameter of the fiber, and fiber type.

There currently exists two analytical methods used to determine airborne asbestos fiber concentrations: phase-contrast microscopy (PCM) and transmission electron microscopy (TEM). PCM is the most commonly used approach. PCM has serious drawbacks, however, as phase-contrast microscopes cannot visualize the fibers with smaller diameters, nor can they differentiate types of fibers. In fact, they can't discern asbestos fibers from cellulose, glass fibers or mineral wool. TEM can visualize all the fibers and can identify the fiber types present. While TEM has other analytical deficiencies, experts concur that it is the best analytical approach commonly available. EPA requires the use of TEM to monitor the air after asbestos removal.

The Harvard Asbestos Symposium concluded, "At the present time there is no single optimum method to characterize potential exposure to airborne fibers from ACM in buildings." The consensus view of the Symposium was that air sampling should be used to "enhance exposure assessment and risk estimation."

NOTIFICATION

Schools - in compliance with AHERA

CONTROL AND ABATEMENT PROCEDURES

Abatement methods are:

1) Operations and maintenance
2) Repair
3) Enclosure
4) Encapsulation
5) Removal

AHERA, which applied to the schools when passed in 1986, may be extended to other facilities. Some people falsely assume that AHERA requires asbestos removal. The law only requires schools to exercise every effort to protect human health and the environment by "the least burdensome method." Financially, the least burdensome method seldom equates with removal.

The threat of lawsuits has prompted building owners to opt for removal; however, removal itself is not without risks. It has the potential to increase rather than decrease indoor air concentrations. Removal and disposal exposes some workers to high concentrations of airborne asbestos. Owners can be legally responsible for the asbestos at the disposal site for 40 years.

Abatement specialists, fully trained with appropriate credentials and experience, should be employed for asbestos detection, encapsulation, removal, disposal and air monitoring procedures. Owners should seek protection by having specialists bonded and carrying insurance.

REFERENCES

Asbestos Abatement Industry Directory, Contact: NICA/NAC, Department 5300, Washington, DC 20061-5030; (703) 683-6422, FAX (703) 549-4838.

Asbestos Removal Reference Directory, Rimbach Publishing Inc., 8650 Babcock Blvd., Pittsburgh, PA 15237.

"Asbestos in Schools - A Special Report," School Business Affairs. December (1988), 26-42.

Berry, G. et al., "Asbestosis: A Study of Dose-Response Relationships in an Asbestos Textile Factory," British Journal of Industrial Medicine, 36 (1980), 98-112.

British Health and Safety Commission. Health and Safety Executive, Asbestos: "Effects on Health of Exposure to Asbestos," Health and Safety Commission, HMSO, London. Guidance note EH.HMSO, 1985.

Commins, B.T., The Significance of Asbestos and Other Mineral Fibers in the Environment. Commins Associates, Pippins, Altwood Close, Maidenhead, Berks S16 4PP, England, 1985.

D'Angelo, W.C., Spicer, R.C.; and Mease, M. "Sophisticated Asbestos Removal: Occupied Buildings and Operating HVAC." Nat. Asbestos Coun. J. 3 (1985), 9-14.

Model EPA Curriculum for Training Management Planners, Environmental Sciences, Ind. and Georgia Institute of Technology. July 1988. Available through ATLIS. 6011Executive Boulevard, Rockville, MD 20852, (301) 468-1916.

National Research Council, Committee on Non-Occupational Health Risks of Asbestiform Fibers. Asbestiform Fibers: Non-occupational Health Risks, National Academy Press, Washington, DC, 1984.

NIBS, Asbestos Model Guide Specifications. NIBS, 1015 15th Street, NW, Washington, DC; (202) 347-5710.

Sawyer, R., "Asbestos Exposure in a Yale Building: Analysis and Resolution," Envir. Res., 13 (1977), 1146-168.

Shanley, Elizabeth and Steven Pike, M.D., "Don't Fall Into the Response Action Trap," School Business Affairs, (December 1988) 23.

Slutsker, Gary, "Paratoxicology" Forbes, January 8, 1990.

Spengler, John D. and Haluk Ozkaynak, John F. McCarthy, Henry Lee, Harvard University, Report Summary of Symposium on Health Aspects of Exposure to Asbestos in Buildings. Conference Proceedings also available.

US Department of Labor, Occupational Safety and Health Administration. 1986. Asbestos Regulations, Title 29, Code of Federal Regulation, 1910.1001 and 1926.58.

US Environmental Protection Agency. Airborne Asbestos Levels in Schools. Washington, D.C., EPA-560/5-81-006, July 1983.

US Environmental Protection Agency. Asbestos Contamination of the Air in Public Buildings. Research Triangle Park, North Carolina, EPA-450/3-76-004, October 1975.

US Environmental Protection Agency. Assessing Asbestos Exposure in Public Buildings: A Report to Congress. Washington, D.C., February 1988.

US Environmental Protection Agency. Guidance for Controlling Asbestos-Containing Materials in Buildings. Office of Pesticides and Toxic Substances, Environmental Protection Agency 560/5-85-024. June 1985.

US Environmental Protection Agency. Measurement of Asbestos Air Pollution Inside Buildings Sprayed with Asbestos. Washington, D.C., EPA-560/13-80-026, August 1980.

Weill, H. and J.M. Hughes, "Asbestos as a Public Health Risk: Disease and Policy," Annual Review of Public Health, 7: (1986), 171-92.

For further information

EPA Asbestos Action Program

OSHA

State agencies of jurisdiction

County health departments

BIOAEROSOLS

DESCRIPTION

Bioaerosols, or airborne biological agents, include allergens and pathogens, such as fungi (yeasts and molds), dander, spores, pollen, insect parts and feces, bacteria and viruses. Bioaerosols are also discussed in literature addressing microbiological, or microbial, contaminants.

Microorganisms associated with normal childhood diseases, such as mumps and measles, or the usual afflictions like flu and "colds" are not treated here.

SOURCES

Sources for biological growth include wet insulation, carpet, ceiling tile, wall coverings, furniture, air conditioners, dehumidifiers, humidifiers, cooling towers, drip pans and cooling coils of air handling units. People, pets, plants, insects may carry biological agents into a facility or serve as potential sources.

Biological agents may enter the building through outside air intakes; and, due to their small size, may not be filtered out of the airstream. Frequently they settle in the ventilation system itself where viable spores and bacteria can incubate and grow. While attempting to curb one contaminant, asbestos, by substituting fiberglass insulation, we have provided a fertile territory for others. The inevitable dust and darkness inside duct work plus condensate moisture, unless carefully maintained, can all work together to turn the vast surface area provided by fiberglass into breeding grounds for mold.

SYMPTOMS AND HEALTH EFFECTS

Common symptoms associated with biological contaminants includes sneezing, watery eyes, coughing, shortness of breath, dizziness, lethargy, fever and digestive problems.

Different causes prompt diverse symptoms and conditions. Concerns may range from odors and "stuffiness" to Legionnaire's disease. No evidence could be found to support the contention that bioaerosols are responsible for problems related to skin or reproductive systems.

Building-related illnesses (BRI) constitute a range of hypersensitivity and infectious diseases. Hypersensitivity diseases, such as asthma, humidifier fever and hypersensitivity pneumonitis, are caused by immunological sensitization to bioaerosols. Prolonged exposure to mold spore allergens, for example, wears down the immune system and increases the individual's sensitivity prompting reactions to lower concentrations. (See Chapter 4 for a discussion of BRI diseases.) In BRI, a physician should be involved in the diagnosis and the etiology.

One case of hypersensitivity pneumonitis is sufficient to suggest the sensitization process is occurring in others, and mitigation procedures should be taken.

LAWS/REGULATIONS/STANDARDS

None at present.

ACCEPTABLE LEVELS

None established.

MONITORING AND MEASUREMENT

Epidemiologists can be very helpful in establishing whether or not a problem exists, and the nature of the problem; and, may be able to hypothesize on the source of the contaminant. Only when positive evidence exists for such diseases as allergic asthma, allergic rhinitis (inflammation of nasal mucous membrane), biological humidifier fever or hypersensitivity pneumonitis, and appropriate maintenance actions do not appear to resolve the problem, should the complex, costly process of bioaerosol sampling be undertaken. The collection process, in general, calls for drawing room air across media. The analysis necessitates an incubation period and the subsequent identification and counting of the colonies found.

If a BRI, such as hypersensitivity pneumonitis, is established, a walk through inspection of the occupied space prior to any complex testing is warranted. Applying the knowledge of the potential sources, the area should be examined; e.g., potential reservoirs, porous wet surfaces, water damaged material, the HVAC system. Presumptive sources for bioaerosols and the nature of the contamination are discussed in relation to maintenance in Chapter 6. This walk through should be accompanied by occupant interviews as discussed in Chapter 5. Remedial action may prove to be very straight forward and testing procedures avoided.

The most helpful document on bioaerosols available is from ACGIH, Guidelines for the Assessment of Bioaerosols in the Indoor Environment (1990). This 87 page pamphlet is comprehensive in scope and is thoroughly referenced.

NOTIFICATION

None required.

CONTROL INTERVENTION OR REMEDIAL PROCEDURES

The list of sources suggests ways to control bioaerosols and microbial contaminants. Most remedies fall in the area of maintenance and preventive maintenance; e.g., cleaning filters and wet areas in the ventilation system; replacing water-damaged carpets, insulation; maintaining relative humidity between 40-60 percent; cleaning/disinfecting drain pans and coils; etc. Chapter 6 discusses bioaerosol-related maintenance. Filtration of bioaerosols is relatively easy as they fall in a size range that can be effectively trapped by a variety of filters. Most microbial cells range from 1-20 μm with a few from 0.5-200 μm and can be removed in a good quality filtration system. 50% atmospheric dust spot efficiency filters will remove most microbial particulates in the return or outdoor airstreams.

ASHRAE 62-1989 has added a discussion of biological contaminants to its earlier 1981 guidelines.

REFERENCES

ACGIH. Guidelines for the Assessment of Bioaerosols in the Indoor Air Environment. Cincinnati: 1990. 87 pages.

Burge, Harriet, "Sample Analysis and Data Analysis," University of Michigan Conference; Assessing Bioaerosol Hazards in the Workplace. September 1987.

Burge, H.A.; Chatigney, M.; Feeley, J.C.; Kreiss, K.; Morey, P.; Otten, J.; and Perterson, K. "Bioaerosols: Guidelines for Assessment and Sampling of Saprophytic Bioaerosolsin the Indoor Environment." Appl Ind Hyg 2(5), 1987.

Morey, P.; Otten, J.; Burge, H.; Chatigny, M.; Feeley, J.; LaForce, M.; and Peterson, K. "Airborne Viable Microorganisms in Office Environments: Sampling Protocol and Analytical Procedures." Appl Ind Hyg 1(1), 1986.

COMBUSTION PRODUCTS

DESCRIPTION

The major categories of products resulting from incomplete combustion can be listed as:

carbon monoxide (CO)

nitrogen oxides (NO_x)

particulate material

polynuclear aromatic hydrocarbons (PAH)

Carbon monoxide (CO) is an odorless, tasteless and colorless gas.

Nitrogen oxides (NO_x) includes nitrogen compounds NO, NO_2, N_2O, OONO, ON(O)O, N_2O_4 and N_2O_5. All are irritant gases, which can impact on human health.

Particulates represents a broad class of chemical and physical particles, including liquid droplets. Combustion conditions can affect the number, particle size and chemical speciation of the particles.

Polynuclear aromatic hydrocarbons (PAH) concentrations are usually low indoors. PAH concerns stem from their potential to act synergistically, antagonistically or in an additive fashion with other contaminants. The chemical composition and concentrations of these compounds vary with combustion conditions.

SOURCES

Combustion products are released under conditions where incomplete combustion can occur, including; wood, gas and coal stoves; unvented kerosene heaters; fireplaces under downdraft conditions; and environmental tobacco smoke. Vehicle exhaust is also a serious concern, particularly from underground or attached garages.

HEALTH EFFECTS

Carbon monoxide (CO) has about 250 times the affinity for hemoglobin than oxygen has. When carboxyhemoglobin (COHb) is formed, it reduces the hemoglobin available to carry oxygen to body tissues. CO, therefore, acts as an asphyxiating agent. Common symptoms are dizziness, dull headache, nausea, ringing in the ears and pounding of the heart. Should CO inhalation induce unconsciousness, damage to the central nervous system, the brain and the circulatory system may occur. Acute exposure can be fatal. Young children and persons with asthma, anemia, heart and hypermetabolic diseases are more susceptible.

The extent to which nitrogen oxides (NO_x) affect human health is unclear. The most information is available about nitrogen dioxide (NO_2). NO_2 symptoms are irritation to eyes, nose and throat, respiratory infections and some lung impairment. Altered lung function and acute respiratory symptoms and illness have been observed in controlled human exposure studies and in epidemiological studies of homes using gas stoves. Studies in the United States and Britain have found that children exposed to elevated levels of NO_2 have twice the incidence of respiratory illness as children not exposed.

Combustion particulates can affect lung function. The smaller respirable particles (μm) present a greater risk as they are taken deeper into the lungs. Particles may serve as carriers of contaminants, such as PAH, or as mechanical irritants that interact with chemical contaminants. (See "Respirable Particulates.")

The health effects of polynuclear aromatic hydrocarbons (PAH) are very difficult to determine or predict. PAH's propensity to act in concert with other contaminants complicates any effort to attribute singular cause and effect. It is known that some PAHs are carcinogens while others exhibit pro- or co-carcinogenic potential.

LAWS/REGULATIONS/STANDARDS

None

ACCEPTABLE LEVELS/GUIDELINES

OSHA guideline for CO recommend 50 parts per million in an eight-hour period. The U.S. National Ambient Air Quality Standard is 40 mg/m^3 for a one-hour period.

MONITORING AND MEASUREMENT

Measurement equipment, such as infrared - radiation absorption and electrochemical instruments, to measure CO do exist, but are relatively expensive. Moderately priced real time measuring devices are also available.

NOTIFICATION

Any employee injury resulting from CO exposure must be reported to the state occupational safety and health agency.

CONTROLS AND REMEDIATION

Maintain, properly adjust and operate all combustion equipment. Vehicular exhaust from garages, loading docks, etc. needs to be carefully managed.

The National Aeronautics and Space Administration (NASA) has found that house plants can serve as a living air cleaner for volatile organic chemicals. Flowering plants like the gerbera daisy and chrysanthemums were found effective in removing benzene; the philodendron, spider plant and golden pothos in removing formaldehyde. Other plants found to be effective air purifiers included English ivy, ficus, Chinese evergreen, bamboo palm, peace lily, mass cane and mother-in-law's tongue. Plant leaves and root-soil also acted as purifiers.

When unusually high levels of combustion contaminants are expected in an area, additional ventilation can be used as a temporary measure.

REFERENCES

CO

Gemelli, F., and Cattani, R. "Carbon Monoxide Poisoning in Childhood." British Medical Journal. 291: (1985), 1197.

Pitkin County Health Dept. "Carbon Monoxide Exposures at a Skating Rink." MMWR 35(27) (1986), 435-441.

Russell, H.L.; Worth, J.A. and Leuchak, W.P. "Carbon Monoxide Intoxication Associated With Use of a Gasoline-Powered Resurfacing Machine at an Ice-Skating Rink." MMWR 33(4) (1984), 49-51.

Sorensen, A.J. "The Importance of Monitoring Carbon Monoxide Levels in Indoor Ice Skating Rinks." College Health 34 (1986), 185-6.

U.S. Environmental Protection Agency. Revised Evaluation of Health Effects Associated with Carbon Monoxide Exposure: An Addendum to the 1979 EPA Air Quality Criteria Document for Carbon Monoxide. EPA-600/8-83-033F, 1984.

NO_2

Ahmed, T.; Marchette, B.; Danta, I.; Birch, S.; Dougherty, R.L.; Schreck, R.; Sackner, M.A. "Effect of 0.1 ppm NO_2 on Bronchial Reactivity in Normals and Subject With Bronchial Asthma." Am. Rev. Dis. 125 (1982), 152.

Hazucha, M.J.; Ginsberg, J.F.; McDonnell, W.F.; Haak, E.O.

Jr.; Pimmel, R.L.; Salaam, S.A.; House, D.E.; Bromberg, P.A. "Effects of 0.1 ppm Nitrogen Dioxide on Airways of Normal and Asthmatic Subjects." Journal Applied Physiology: Respir. Environ. Exercise Physiol. 54 (1983), 730-739.

Kleinman, M.T. Bailey, R.M.; Linn, W.S.; Anderson, K.R.; Whynot, J.D.; Shamoo, D.A.; Hackney, J.D. "Effects of 0.2 ppm Nitrogen Dioxide on Pulmonary Function and Response to Bronchoprovocation in Asthmatics." Journal Toxicology Environ. Health 12 (1983), 815-826.

Kuraitis, K.; Richters, A.; Sherwin, R.P. "Spleen Changes in Animals Inhaling Ambient Levels of Nitrogen Dioxide." Journal Toxicology Environ. Health 7 (1981), 851-859.

McGrath, J.J.; Oyervides, J. "Resyponse of NO_2-Exposed Mice to Klebsiella Challenge." In: International Symposium on the Biomedical Effects of Ozone and Related Photochemical Oxidants. Lee, S.D.; Mustafa, M.G.; Mehlman, M.A. (Eds.). Princeton, N.J.: Princeton Scientific; (1983), 475-485.

McGrath, J.J.; Smith, D.L. "Respiratory Responses to Nitrogen Dioxide Inhalation." Journal of Environ. Sci. Health Part A A19 (1984), 417-431.

Richters, A.; Kuraitis, K. "Inhalation of NO_2 and Blood-Borne Cancer Cell Spread to the Lungs. "Arch. Environ. Health 36 (1981), 36-69.

For further information

State environmental safety and health agencies

County health department

ENVIRONMENTAL TOBACCO SMOKE

DESCRIPTION

Environmental Tobacco Smoke (ETS) comes from the sidestream smoke emitted from the burning end of cigarettes, cigars and pipes and secondhand smoke exhaled by smokers.

Breathing in ETS is generally referred to as passive, or involuntary, smoking.

ETS contains a mixture of irritating gases and carcinogenic tar particles. Because tobacco doesn't burn completely, other contaminants are given off, including: sulfur dioxide, ammonia, nitrogen oxides, vinyl chloride, hydrogen cyanide, formaldehyde, radionuclides, benzene and arsenic.

SOURCES

A lighted cigarette gives off approximately 4,700 chemical compounds. The EPA has estimated that 467,000 tons of tobacco are burned indoors each year.

Benzene, a known human carcinogen, is emitted from synthetic fibers, plastics and some cleaning solutions; however, the most important source of exposure is cigarettes. Benzene levels have been found to be 30 to 50 percent higher in homes with smokers than in nonsmoking households.

SYMPTOMS AND HEALTH EFFECTS

Carbon monoxide, nicotine and tar particles have been identified as the chemicals most apt to impact on health.

In 1985, three federal bodies independently arrived at the same conclusion: passive smoking significantly increases the risk of lung cancer in adults. The Surgeon General warned: "A substantial number of the lung cancer deaths that occur among nonsmokers can be attributed to involuntary smoking."

According to the EPA, tobacco smoke contains 43 carcinogenic compounds. The agency also notes that ETS is a major source of mutagenic substance; i.e., compounds that cause permanent changes in the genetic material of cells.

Studies have shown that passive smoking significantly increases respiratory illness in children. Asthmatic children are particularly at risk.

The federal Interagency Task Force on Environmental Cancer, Heart and Lung Disease Workshop on ETS concluded that the effects of ETS on the heart may be an even greater concern than its effect on the lungs. Several studies have linked passive smoking with heart disease.

LAWS/REGULATIONS/STANDARDS

Federal laws prohibit the sale of cigarettes to persons sixteen years of age or younger and forbid smoking on any domestic commercial flight.

Many states and local ordinances prohibit smoking in designated facilities or areas. Some require businesses serving the public, such as restaurants, to provide non-smoking areas.

ASHRAE 62-1989 recommends 60 cfm/person for smoking lounges.

ACCEPTABLE LEVELS

There is no established health-based threshold for ETS exposure. Since it is recognized as a cancer causing agent, the EPA recommends that ETS exposure be minimized wherever possible.

Cain found 90 percent occupant satisfaction with 2 ppm CO as produced by cigarette smoking. His research also revealed 780 ft^3 of ventilation air per cigarette would satisfy this level. On this basis, two cigarettes per hour by 30 percent of the occupants = 8 cfm/person.

$$\text{cfm/person} = \frac{780\ ft^3/\text{cig} \times 2\ \text{cig/hr} \times 30\%}{60\ \text{min/hr} \times 100}$$

Since ETS is composed of many contaminants, an additional reference point is to consider acceptable levels of some of the major ETS contaminants. The table on the following page offers the current threshold limit values (TLVs) recommended by the American Conference of Governmental Industrial Hygienists for exposure in the workplace for CO, nicotine, NH_3, HCN and phenol.

MONITORING AND MEASUREMENT

The level of involuntary ETS exposure is hard to establish due to: (1) the complexity of smoke composition; (2) variability of the ETS composition;

and (3) ETS contaminants' similarity to other airborne pollutants. ETS exhibits both spatial and temporal variations. Over time the particulate size changes and volatile components are lost from ETS particles. In addition, conditions change during the puff and smolder periods.

While far from perfect, visibility and odor do provide indicators of exposure level. Using odor as an indicator can be deceptive, however, as perceived odor intensity decreases rapidly with time of exposure, while the irritation actually increases during the first hour of exposure and then tends to achieve a steady state thereafter (Clausen et. al., 1987).

Wheeler (1988) and Cain et. al. (1987) have found that carbon monoxide (CO) concentration is the most direct index of ETS and a better measure of acceptability than cfm/person. Nystrom et. al. (1986) did note some problems with using CO, observing that CO from ETS is generally less than 1 ppm above CO background levels. CO from heating, cooking and car exhaust makes it difficult to show the CO link to ETS except in chamber studies.

TABLE A-1

EFFECT OF DILUTION ON CONCENTRATION OF INDIVIDUAL ETS COMPONENTS ASSUMING "TAR" AS REFERENCE MATERIAL AND COMPARISON TO TLVS (ACGIH 1984-85)

SMOKE COMPONENT	TYPICAL SIDE STREAM YIELD	CONCENTRATION $\mu g/m^3$		TLV-TWA $\mu g/m^3$
'TAR'	14.9 mg	25	250	—
CO	58.0 mg	97.9	979	55000
NICOTINE	4.39 mg	7.4	74	500
NH_3	6.75 mg	11.3	113	18000
HCN	71 μg	0.12	1.2	10000(c)
PHENOL	68 μg	0.11	1.1	19000
CATECHOL*	83.5 μg	0.14	1.4	
BAP*	52.5 μg	0.000087**	0.00087	
NDMA*	657 μg	0.00110	0.0110	

* Based on values reported by Hoffman et al. (1985).

** This is 87 picograms.

NOTIFICATION

Local ordinances may ban smoking in some locations, such as airports, and may require the designation and posting of nonsmoking areas in facilities serving the public.

Company policies regarding nonsmoking/smoking in designated areas should be circulated and the designated nonsmoking areas posted.

CONTROL AND REMEDIAL PROCEDURES

The control and treatment of ETS generally falls into five areas:

(1) remove the source - eliminate smoking. Decrees to eliminate smoking are policy decisions with employee relations considerations.

(2) modify the source - relocate/separate smokers. Separating smokers will reduce, but not eliminate, ETS. (Refer to chapter 3 for management considerations.)

(3) dilution ventilation. Increasing ventilation can prove to be an expensive option. Ventilation rates to satisfy ETS - related health concerns have not been established. ASHRAE 62-1989 guidelines are designed to reduce tobacco smoke odor; not necessarily reduce health risks. (See discussion Chapter 10, ASHRAE 62-89.)

(4) filter the contaminants. HEPA filters and electrostatic precipitators can remove respirable particles. Since both odor and irritation are mainly caused by the gaseous phase of smoke, this is not a particularly viable option. Granulated filter media, such as activated carbon, or some type of catalytic system can collect the volatile components of ETS. EPA has cautioned that the use of filters to remove ETS is technically and economically impractical. House plants have also been shown to be helpful in removing contaminants. (See sections on "Combustion Contaminants" and "Formaldehyde.")

(5) isolate smokers and exhaust contaminants to outside. Establish a smoking lounge, where the return air ductwork is blocked and exhaust registers move air directly outside.

REFERENCES

Adams, J.D.; O'Mara-Adams, K.J.; Hoffmann, D. "On the Mainstream-Sidestream Distribution of Smoke Components from Commercial Cigarettes." 39th Tobacco Chemists' Research Conference. Montreal, PQ. 1985.

Cain, W.S.; Tosun, T.; See, L.; and Leaderer, B.P. "Environment Tobacco Smoke: Sensory Reactions of Occupants." Atmospheric Environment, 21, 2 (1987) 347-357.

Cain, W.S.; Leaderer, B.P.; Isseroff, R.; Berglund, L.G.; Huey, R.J.; Lipsitt, E.D.; and Perlman, D. "Ventilation Requirements in Buildings-I. Control of Occupancy Odor and Tobacco Smoke Odor." Atmosphere Environment, Vol. 17, 6 (1988), 1183-1197.

Clausen, G.H.; Moller, S.B.; Fanger, P.O.; Leaderer, B.P.; and Dietz, R. "Background Odor Caused by Previous Tobacco Smoking." Managing Indoor Air for Health and Energy Conservation, Atlanta: ASHRAE, (1986), 119-125.

Clausen, G.H.; Moller, S.B.; Fanger, P.O. "The Impact of Air Washing on Environmental Tobacco Smoke Odor." In B. Seifert et al. (eds.): Proceedings of Indoor Air '87, Berlin, Vol.2, (1987), 47-51.

Committee on Passive Smoking. National Research Council, National Academy of Sciences. Environmental Tobacco Smoke Measuring Exposures and Assessing Health Effects. (1986), 245.

First, M. "Environmental Tobacco Smoke Measurements: retrospect and prospect." European Journal of Respiratory Disease. Vol. 65 (Supplement), (1983), 9-16.

Ingebrethsen, B.J., and Sears, S.B. "Particle Size Distribution Measurements of Sidestream Cigarette Smoke." Proceedings of the Tobacco Chemists' Research Conference. Montreal, Canada, October, 1985.

Leaderer, B.P. and W.S. Cain, "Air Quality in Buildings during Smoking and Non-Smoking Occupancy." ASHRAE Trans. V89 Part A & B, (1983), Paper No. DC-83-11.

Muramatsu, T.; Weber, A.; Muramatsu, S.; and Akerman, F. "An Experimental Study on irritation and Annoyance Due to Passive Smoking." Int. Arch. Occup. Environ. Health, Vol. 51 (1983), 305-17.

Repace, J.L. "A Quantitative Estimate of Nonsmokers' Lung Cancer Risk from Passive Smoking," Environment Int., Vol. 11 (1985), 3.

Spengler, J.D., and Soczek, M.L. 1984. "Evidence for Health Effects of Side-Stream Tobacco Smoke." ASHRAE Transactions. Vol. 84, Part 15 (1984), 781-788.

Sterling, T.D. and Sterling, E.M. "Investigations on the Effect of Regulating Smoking on Levels of Indoor Pollution and on the Perception of Health and Comfort of Office Workers." Eur J. Respir Dis, Suppl 133, Vol, 65 (1984), 17-32.

U.S. Public Health Service. "The Health Consequences of Involuntary Smoking." Surgeon General's Report. (1986), 332.

William S. Cain and Brian P. Leaderer. "Ventilation Requirements in Occupied Spaces During Smoking and Nonsmoking Occupancy." Environment International, Vol. 8. (1982), 505-514.

For further information

Office on Smoking and Health
U.S. Public Health Service
5600 Fishers Lane, Room 1-10
Rockville, MD 20857

American Society of Heating
Refrigerating and Air Conditioning Engineers (ASHRAE)
1791 Tullie Circle, NE
Atlanta, GA 30329

Office of Cancer Communications
National Cancer Institute
1-800-CANCER

Public Information Center
U.S. Environmental Protection Agency
Mail Code PM-211B
401 M Street, SW
Washington, DC 20460

FORMALDEHYDE

DESCRIPTION

Formaldehyde (HCHO) is a chemical, a volatile organic compound. It is used in a wide variety of products and is most frequently introduced into the building during initial construction or renovation. It is a colorless gas at room temperature and has a pungent odor.

Formaldehyde is a ubiquitous chemical influenced by temperature and humidity. Concentrations tend to be highest in prefabricated homes, ranging from 0.1 to 5.0 mg/m^3 (1 mg/m^3 = 0.8 13 ppm HCHO). More commonly the range is 0.1 to 1.0 mg/m3.

SOURCES

Formaldehyde is used in many building products, including plywood, paneling, particleboard, fiberboard, urea-formaldehyde foam insulation, adhesives, fiberglass and wallboard. Other potential sources are furniture, shelving, partitions, ceiling tiles, wall coverings, draperies, upholstery, carpet backing and ceiling tiles.

Formaldehyde is also a product of incomplete combustion; therefore, cigarette smoke, and cooking/heating fuels, such as natural gas and kerosene are sources.

SYMPTOMS AND HEALTH EFFECTS

Clinical and epidemiological data indicate human response to formaldehyde can vary greatly. Some people exhibit hypersensitive reactions. Acute exposure to formaldehyde can result in eye, ear, nose and throat irritation, coughing and wheezing, fatigue, skin rash and severe allergic reactions. It is a highly reactive chemical that combines with protein and can cause allergic contact dermatitis. Table A-2 shows the effect of short term exposure with ranges and median responses, excluding immunosensitive populations.

TABLE A-2. EFFECT OF FORMALDEHYDE IN HUMANS AFTER SHORT-TERM HCHO RESPIRATORY EXPOSURE

REPORTED RANGES*	ESTIMATED MEDIAN*	EFFECT
0.06-1.2	0.1	Odor threshold 50% of people
0.01-1.9	0.5	Eye irritation threshold
0.1-3.1	0.6	Throat irritation threshold
2.5-3.7	3.1	Biting sensation in nose, eye
5-6.7	5.6	Tearing eyes, long term lung effects
12-25	17.8	Tolerable for 30 minutes with strong flow of tears lasting 1 hour
37-60	37.5	Inflammation of lung (pneumonitis), edema, respiratory distress: danger to life
60-125	—	Death

* Concentrations in mg/m^3; 1 mg/m^3=0.813 ppm
Source: National Center for Toxicological Research (1984)

Controversy remains regarding the carcinogenic role of formaldehyde in humans (Sun 1986). The incidence of cancer in rodents makes formaldehyde highly suspect as a human carcinogen. EPA has recently conducted research which suggests formaldehyde may cause a rare form of throat cancer in long-term occupants of mobile homes. Chamber studies (Cain 86) have shown that a given concentration of formaldehyde may evoke quite different degrees of irritation, depending upon duration of exposure, fluctuations in concentrations and the presence of other agents in the air.

LAWS/REGULATION/STANDARDS

A. ACGIH

1. TLV - 1 ppm (1.5 mg/m$^{3)}$
2. STEL - 2 ppm (15 mins.)
3. Notation - Suspect Human Carcinogen

B. AIHA Community Air Quality Guideline

1. Short Term Exposure Level (30 mins.)
2. Exposure Level - 120 μg/m^3

C. ASHRAE Selected Guidelines for Air Contaminants of Indoor Origin

1. Continuous
2. Exposure Level 120 μg/m^3

D. HUD - 24 CFR 3280

1. Applies to Mobile Homes
2. Standard sets emission rates and test conditions

E. OSHA (MOSH) - 29 CFR 1910.1048: Adopted in 1988

1. PEL - 1 ppm (Action Level 0.5 ppm)
2. Many requirements for occupational exposures, including:
   a) Initial Monitoring
   b) Housekeeping (spills) Procedure
   c) Training

ACCEPTABLE LEVELS AND BACKGROUND INFORMATION

Guidelines, in Section 3 above, set suggested and required levels and actions under various conditions.

Indoor levels of 0.1 ppm are considered problematic. The Health and Welfare Canada guideline is 0.1 ppm. Studies by the United States government (CPSC and HUD) have shown that typical formaldehyde levels range from .01 ppm to 1.0 ppm depending on the type of structure and materials used. Homes, older than 5 years without urea-formaldehyde foam insulation, range from .02 ppm to .05 ppm (Hawthorne et. al., 85). Office buildings range from .01 to .08 ppm.

MONITORING AND MEASUREMENT

Formaldehyde is found in virtually all indoor environments and is routinely monitored with active and/or passive devices in almost all IAQ diagnostic investigations.

The presence of new construction, modifications, refurbishments, furniture and the presence of typical symptoms is cause for measuring formaldehyde levels. While a probable cause, such as new carpeting or drapes, may be localized and obvious; in other instances, sources may be more subtle, such as replaced ceiling tiles or chip board shelving. Ventilation systems could prompt widespread complaints away from the source.

Detector tubes are the most commonly employed method of assessing formaldehyde levels, allowing measurements of 0.2-5.0 ppm (varies with manufacturer). Readings can be made in minutes.

Samples over a longer period, such as one week, can be made with badges or passive diffusion tubes. These dosimeters and the badges rely on a color change principle and can be read immediately. However, they offer a relatively low degree of accuracy, as other gases may influence the reading.

Most passive monitoring devices are generally not sensitive enough to detect formaldehyde below 0.1 ppm in 4 or 8 hour sampling periods. A "passive bubbler" now available from Air Technology Labs (ATL) can measure formaldehyde at indoor air concentrations of less than 0.1 ppm. Sampling time of two to four hours is usually sufficient for the ATL device. For an 8 hour sampling time, the lower quantification limit is 0.025 ppm

(0.03 mg/m^3). The results can be analyzed in the field in less than half an hour after sampling. Reliable results require some air movement.

NOTIFICATION

None in non-industrial situations.

CONTROL AND REMEDIAL PROCEDURES

Through selective purchasing, materials can be obtained with lower potential formaldehyde off-gassing. Researchers have found up to a 23 fold difference in emission from the same products from different manufacturers. Goods can also be pre-treated to reduce emissions levels. Contact and inquiries with dealers and manufacturers with regard to formaldehyde can result in materials with lower exposure levels.

The potential to significantly reduce the irritation from formaldehyde warrants a pro-active, preventive approach. Laminated products should have all exposed surfaces or unused "plug holes" covered with laminate or replugged.

Pretreating carpet by heating or steaming can accelerate off-gassing. Similarly, "bake-out" procedures may accelerate off-gassing from other construction materials and furnishings. Purging with outside air after a bake-out, with the return air blocked, can air out the facility before occupancy.

Increased ventilation rates within the typical indoor parameters have not shown any significant reduction in formaldehyde concentrations in several studies. While other studies have shown some improvement, ventilation is not the preferred control method.

Barrier coatings and sealants might be used to reduce formaldehyde emissions. Researchers at Ball State University (Godish) have tested various sealants on particleboard flooring and have found they are effective in the short run and after six months. Barrier coatings and sealants pose their own IAQ problems and adequate ventilation should be maintained during application and until the strong odor fades. Prior to using sealants, notification to chemically sensitized people is recommended.

A study by the National Aeronautics and Space Administration (NASA) has found ordinary house plants can significantly reduce levels of formaldehyde. NASA found philodendron, spider plant and golden pothos were most effective. The agency also determined that the root-soil zone was a vital part of removing contaminants and recommended maximizing air exposure to the plant root-soil area. Activated carbon filters containing fans as an integrated part of any plan using houseplants for large volume areas was suggested.

REFERENCES

Air Quality Research International, "Performance of PF-1 and Dupont Passive Formaldehyde Monitors in AQRI Test Atmosphere Exposure Chamber." Prepared for the UFFI Center, 1984.

Air Technology Labs, 548 E. Mallard Circle, Fresno, CA 93710 (209) 435-3545.

Balmat, J.L. "Generation of Constant Formaldehyde Levels for Inhalation Studies." American Industrial Hygiene Association Journal, Vol. 46(6) (1985), 690-692.

Balmat, J.L., and Meadows, G.W. "Monitoring of Formaldehyde in Air." American Industrial Hygiene Association Journal, Vol. 46(10) (1985), 578-584.

Bender, J.R.; Mullin, L.S.; Graepel, G.J.; and Wilson, W.S. "Eye Irritation Response of Humans to Formaldehyde." American Industrial Hygiene Association Journal, Vol. 44(6) (1983) 463-465.

Black, M. "Formaldehyde in Construction and Building Materials." Georgia Institute of Technology, Indoor Air Quality Symposium, Atlanta, GA. 1985.

Black, M., et. al. "Correlation of Wood Product Formaldehyde Emission Rates as Determined Using a Large Scale Test Chamber, Small Scale Test Chamber, and Formaldehyde Emission Monitor," Georgia Institute of Technology, Atlanta, GA. 1985.

Cain, W.S. "Sensory Attributes of Cigarette Smoking." Banbury Report No. 3: A Safe Cigarette? Cold Spring Harbor Lab, (1980), 239-249.

Cain, W.S., and Murphy C.L. "Interaction Between Chemoreceptive Modalities of Odour and Irritation." Nature, Vol. 284, (1980), 255-257.

Godish, Thad, Ball State University, Muncie, Indiana.

Godish, Thad and Jerome Rouse, "Control of Residential Formaldehyde Levels by Source Treatment." Indoor Air '87 Proceedings, 4th International Conference on Indoor Air Quality and Climate, Vol. 3. Berlin, West Germany, (August, 1987), 221-225.

Grot, R. et. al. "Validation of Models for Predicting Formaldehyde Concentrations in Residence Due to Pressed Wood Products," NBS, NBSIRT85-3255, Gaithersburg, MD. 1985.

Hanrahan, L.P.; Anderson, H.A.; Dally, K.A.; Eckmann, A.D.; and Kanerek, M.S. "Formaldehyde Concentrations in Wisconsin Mobile Homes." Journal of the Air Pollution Control Association, Vol. 35,(1985), 1164-1167.

Meyer, B., and Hermanns, K. "Reducing Indoor Air Formaldehyde Concentrations." Journal of the Air Pollution Control Association, Vol. 35, (1985), 816-821.

U.S. Housing & Urban Development "Air Chamber Test Method for Certification and Qualification of Formaldehyde Emission Levels," Federal Register, 49, 155, 24 CFR 3280.406.

U.S. Housing & Urban Development "Evaluation of Formaldehyde Problems in Residential Mobile Homes." Technology and Economics, Government Printing Office, Washington, D.C. 1981.

For further information

Contact state agencies with OSHA or environmental responsibilities

RADON

DESCRIPTION

Radon is an odorless, colorless gas that is always present at various concentrations in the air. Radon is formed from the decay of radium, which in turn results from the decay of uranium. Radon (Rn-222) and more specifically, the radio active elements to which it decays (radon daughters) represent a major indoor air concern, particularly in homes.

Radon daughters are charged particles and they, in turn, absorb particles in the air. Approximately 90 percent of the radon daughters attach to airborne particles before they are inhaled. The remaining 10 percent represent a significant source of exposure; for, as smaller particles, they deliver a dose to critical lung cells. (See following section, Respirable Particulates.) About 30 percent of the inhaled daughters are deposited in the lung.

Radon has a radiological half-life of 3.8 days, while the daughters' half-life is about 30 minutes. This rapid decay means the daughters emit high radiation energy levels to a comparatively small volume of tissue; and, in the process, provide a major source of injury to tissues.

Radon exhibits daily and seasonal variation in concentration. Fleming and others have found extreme variability (7.5 to 25 pCi/l) in a diurnal cycles in many buildings that do not coincide with ventilation rates. Indoor radon concentrations in most climates are much lower in the summer. This is usually attributed to more open windows and doors, which increases ventilation and tends to equalize pressure differentials.

SOURCES

Some radon will enter a facility through the water system and off-gassing of building materials; however, the principal source of radon is the soil. Radon typically enters through cracks, voids or other openings in the foundation.

Conditions affecting the flow of radon into a facility are;

soil factors - level of radon concentration, emanation rate, diffusion length, permeability, soil moisture.

building factors - type and formation of the foundation or substructure, construction quality, design

pressure differentials - building depressurization through stack (thermal) effect and/or exhaust fans, wind, barometric pressure changes, pressure gradients in the soil.

SYMPTOMS AND HEALTH EFFECTS

No immediate symptoms are associated with radon exposure.

When absorbed into the lung cavity, radon decay products may increase the incidence of lung cancer. There is evidence that tobacco smokers exposed to radon are more likely to get lung cancer.

LAWS/REGULATIONS/STANDARDS

Federal, state and local laws with regard to permissable radon levels, testing, training and licensing of mitigators, notification procedures and real estate transactions are changing rapidly. For current provisions, contact local government, state agencies with health and environment jurisdiction, and the EPA, Radon Division.

ACCEPTABLE LEVELS

EPA has recommended 4 picocuries per liter (pCi/l) as an action level for homes. This recommended level is being accepted in the promulgation of laws and ordinances. Parsippany - Troy Hills, New Jersey's ordinance, for example, requires commercial and public buildings be tested and all facilities over 4 pCi/l be mitigated to below 4 pCi/l within 6 months.

An Austrian research study by Fleck (1988) found reduced incidence of chemically induced lung cancer at radiation exposure levels equivalent to continuous exposure at 4 pCi/l.

Working level (WL) is used as a unit of radon exposure, a weighted average of the concentration of radon daughters. The working level month (WLM) is a unit of cumulative exposure with 1 WLM as 1 WL for an occupational month of 170 hours. Radon exposure of 0.02 WL are suggested maximum levels.

MONITORING AND MEASUREMENT

Because of the extreme diurnal and seasonal variations in radon concentrations, measuring and monitoring radon levels is difficult.

Two common forms of testing are the charcoal canister and the alpha tracker. Readings from the canister represent only a "snapshot" in time and do not allow for variations in radon concentrations. Canister results are frequently used to determine if testing using an alpha tracker is warranted. No mitigation should be considered without alpha tracker information.

Verification of radon testing equipment's reliability and the analyst's capabilities prior to use is recommended. Some states require certification of testers. EPA has standards for testing, mitigation and training.

Sample testing needs to be done in several locations in a facility as readings vary greatly from room to room. A number of basement and first floor locations should be tested and analyzed to determine if testing upper floors is needed.

NOTIFICATION

Notification of testing, test results and/or mitigation plans vary. Contact local, state agencies.

CONTROL AND REMEDIATION PROCEDURES

The primary sources of radon in buildings with high concentrations is the pressure driven flow of radon soil gas. This pressure difference may be the result of indoor-outdoor temperature differences, prevailing winds, the mechanical ventilation systems and combustion devices that have a depressurizing effect on the building.

The most effective means of controlling radon is to prevent it from entering the building.

In residences, mechanical ventilation of crawl spaces has been very effective.

Using fans to draw soil gas from beneath a slab in sub-slab ventilation has been shown to be 50 to 90 percent effective in reducing radon levels.

In situations where a basement is fairly tight, pressurizing the basement has been very effective. This relatively simple and inexpensive process has reduced radon concentrations to 65-95 percent below radon concentration guidelines.

Facilities with concrete block walls offer many, many avenues for radon entry. Sealing the block walls and ventilating the cavities have cut radon levels up to 90 percent. While sealing all cracks and openings around drains is a good idea, it has actually met with mixed success.

Provisions to increase ventilation at the first floor level and exhaust out of the basement as conducted in a Pennsylvania study (Wellford) proved to be effective as source control using positive pressure remedies.

Increased ventilation is frequently recommended in the literature as a way to reduce indoor air concentrations. Repeated studies have found no correlation between radon concentrations and air exchange rates. (See Figure 1-1.)

Precautions should be taken in new construction to thwart radon entry. Recommendations made by Brennan and Turner several years ago in "Defining Radon" still provided excellent control guidance. They recommend: (1) slabs and basements be poured with as few joints as possible and poured right to the wall; (2) wire reinforcement be used in slabs and walls help prevent future cracks; (3) seams and perimeters be caulked with polyurethane; (4) dampproofing, sealing and coating the walls be done to slow entry; and (5) sub-slab drains and several inches of #2 stone be placed below the slab during construction to provide sub-slab ventilation potential.

Several states require that those who engage in, or profess to engage in, testing for radon gas or the abatement of radon gas be certified or licensed. The state environmental agency can provide information as to whether such credentials are required; and, if so, provide lists of those who meet the requirements.

REFERENCES

Health Risks of Radon, National Academy Press, 2101 Constitution Avenue, N.W., Washington, D.C. 20418

National Council on Radiation Protection and Measurements. Evaluation of Occupational and Environmental Exposures to Radon and Radon Daughters in the United States. Bethesda, MD: National Council on Radiation Protection and Measurements: NCRP Report No. 78, 1984.

National Council on Radiation Protection and Measurements. Exposures from the Uranium Series with Emphasis on Radon and Its Daughters. Bethesda, MD: National Council on Radiation protection and Measurements: NCRP Report No. 77, 1984.

Radon Epidemiology: A Guide to the Literature, Susan L. Rose, Program Manager, ER-73, Office of Health and Environment Research, Department of Energy, GTN, Washington, D.C.20545; (301) 353-4731.

Straden, E. "Radon in Dwellings and Lung Cancer -- a Discussion." Health Physics 38 (1980), 301-306.

The Radon Industry Directory, Radon Press, 500 N. Washington St., Alexandria, VA 22314.

U.S. Department of Energy. Indoor Air Quality Environmental Information Handbook: Radon. Washington, D.C.: U.S. Department of Energy. DOE/PE/72013-2, 1986.

U.S. Environmental Protection Agency. A Citizen's Guide to Radon, 1986.

For further information

Local government, county health department

State agencies with health and environment jurisdictions

U.S. Department of Health and Human Services

U.S. Environmental Protection Agency
EPA radon hot line: 1-800-SOS-RADON (Yes, that's one too many digits, but it works.)

North American Radon Association

American Association of Radon Scientists and Technologists

RESPIRABLE PARTICULATES

DESCRIPTION

Particulates represent a broad class of chemical and physical contaminants found in the air as discrete particles. Respirable particulates are generally defined as 10 microns or less in size. They fall into two categories; biological and nonbiological particles include plant and animal matter as well as micorbial particles. Size also determines the magnitude of risk and the ultimate location. Since smaller particles are breathed deeper into the lungs, they can by-pass respiratory defense mechanisms thereby creating a greater health hazard. Smaller particles also offer a greater surface area to total mass ratio. Since respirable particles also serve as the carriers for other contaminants, such as PAH (See "Combustion Contaminants.") and pathogens, they can deliver harmful substances to more vulnerable areas.

Particles associated with particular contaminants, such as asbestos and bioaerosols, are treated in separate sections of this appendix.

SOURCE

Minor sources are office equipment, skin particles, occupant exhalants and biological agents. Radon and radon daughters also contribute. Common sources of respirable particulates include kerosene heaters, humidifiers, wood stoves and fireplaces. The largest single contributor to indoor air particle concentrations, particularly in an office environment, is environmental tobacco smoke.

SYMPTOMS AND HEALTH EFFECTS

Particulates cause a wide range of health and comfort problems, which varies with size and type of particles. Irritation and infections in respiratory track and eye irritation are all symptoms associated with respirable particles. All symptoms and health effects associated with environmental tobacco smoke also apply. Respirable particulates are also associated with lung cancer.

Amman et. al. (1986) list the following concerns related to respirable particles;

(1) chemical or mechanical irritation of tissues, including nerve endings at the site of deposition,

(2) impairment of respiratory mechanics,

(3) aggravation of existing respiratory or cardiovascular diseases,

(4) reduction in particle clearance and other host defense mechanisms,

(5) impact on host immune system,

(6) morphologic changes of lung tissues, and

(7) carcinogenesis.

Health consequences vary with the size, mass, concentration and other contaminants acting in concert with the particles. EPA (1981, 1984) has found that respirable particles at concentrations of 250 to 350 $\mu g/m^3$ increase respiratory symptoms in compromised individuals.

LAWS/REGULATIONS/STANDARDS

None

MONITORING AND MEASUREMENT

Refer to particular contaminant discussions, such as ETS.

NOTIFICATION

None

CONTROL AND REMEDIAL PROCEDURES

Particulate control strategies include dilutions with outdoor air, when indoor air particulate concentrations are greater than outside. This, however, is a relatively expensive option.

A second control strategy is recirculating indoor air, which allows mixing with some outside air and/or reconditioning within the ventilation system. Recirculated air must be diluted or treated, or particle concentrations will increase as long as the source is producing.

Particles are the easiest contaminants to remove from the air stream. Media filters and electrostatic air cleaners are used in building HVAC systems. High efficiency particulate air (HEPA) filters and electrostatic precipitators can remove respirable particulates from the air stream very efficiently.

Appropriate filter maintenance with scheduled cleaning and replacement is required. (See Chapter 6, Filters, for further information.)

REFERENCES

Amman, H.M., and M.A. Berry, N.E. Childs D.T. Mage. "Health Effects Associated with Indoor Pollutants," Managing Indoor Air for Health and Energy Conservation. Atlanta: ASHRAE,1986.

Chang, P., Peters, L.K., and Ueno, Y. "Ventilation Requirements in Occupied Spaces During Smoking and Nonsmoking Occupancy." Environmental International, Vol. 8. (1985), 505-514.

Leaderer, B.P., Cain, W.S., Isseroff, R, and Berglund, L.G. "Ventilation Requirments in Buildings - II. Particulate Matter and Carbon Monoxide from Cigarette Smoking." Atmospheric Environment, Vol. 18, No. 1 1984), 99-106.

Meckler, M. and J.E. Janssen, "Use of Air Cleaners to Reduce Outside Air Requirements," Engineering Solutions to Indoor Air Problems. (1988), 130-147.

Owen, M.K. and D.S. Ensor, L.S. Hovis and W.G. Tucker. "Effects of Office Building Heating and Ventialtion System Parameters on Respirable Particles," Managing Indoor Air for Health and Energy Conservation. (1986), 510-516.

Wheeler, Arthur E. "Office Building Air Conditioning to Meet Proposed ASHRAE Standard 62-1981 R." Engineering Solutions to Indoor Air Problems. (1988), 99-107.

VOLATILE ORGANIC COMPOUNDS (VOCS)

DESCRIPTION

Organic compounds that exist as a gas or can easily off-gas under normal room temperatures and relative humidity are considered volatile. More precisely, if an organic compound has vapor pressures greater than 0.1 mm Hg at 20°C, it is considered volatile.

NOTE: Formaldehyde is a VOC. Tobacco smoke and other combustion contaminants emit VOCs such as formaldehyde, benzene, phenols. For more specific treatment of formaldehyde, ETS or combustion contaminants, please see other sections in Appendix A.

SOURCES

Hundreds of VOCs are found in the indoor air. The list of potential sources is lengthy. Some of the major, and more common sources, are photocopying materials, paints, gasoline, people, refrigerants, personal hygiene and cosmetic products, building materials, molded plastic containers, disinfectants, cleaning products and environmental tobacco smoke. Some of these sources emit several VOCs. For example, tobacco smoke contains such VOCs as alcohols, acetone, benzene, formaldehyde, phenols, ammonia, aromatic hydrocarbons and toluene. Few are unique to any one source, as toluene, for instance, can also be found in gasoline, paints, adhesives and solvents.

SYMPTOMS AND HEALTH EFFECTS

Symptoms attributable to VOCs include respiratory distress, sore throat, eye irritation, nausea, drowsiness, fatigue, headaches and general malaise.

Specific VOCs are not often proven to cause SBS complaints. Due to the large numbers of chemicals found indoors, it is very difficult to establish any causal relationship between health and certain VOCs. Occupational exposure studies have documented respiratory ailments, heart disease, allergic reactions, mutagenicity and cancer to some VOCs. Combinations of certain VOCs are suspected of having synergistic effects and this potential is currently being researched.

LAWS/REGULATIONS/STANDARDS

None exist for non-industrial settings.

ACCEPTABLE LEVELS

Dose-response information is insufficient to establish an association between measured air concentrations of VOCs and specific complaints. In a few instances, cause effect relationships have been made. Molhave's (1984) work suggested that a mixture of 22 VOCs at levels of 5 to 25 mg/mm^3 had acute neurological effects in the Danish buildings study.

OSHA has Threshold Limit Value (TLV) standards for industrial work spaces. ASHRAE 62 recommends VOC concentration levels of OSHA TLVs.

Thresholds generally relate to single compounds. TLV's do not take into account the effects of simultaneous or serial exposure to complex mixtures. European researchers have shown that exposure to a mixture of VOCs typically found indoors will cause eye and respiratory irritations, headaches, general malaise, etc. at concentrations of 8 - 25 mg/m^3 with no observable effects below 3 mg/m^3.

MONITORING AND MEASUREMENT

Actual measurements for VOCs, including formaldehyde, for orientation purposes should not be made in the early investigation stage except in new or refurbished buildings where VOCs are suspected.

VOCs have been monitored using real-time measurements and solid sorbents collective followed by thermal desorption gas chromatography/mass spectrometry (GC/MS) analysis. Inexpensive monitors suitable for work in the field exist, but they are not particularly reliable.

Drawing on the work of the World Health Organization and refined by Indoor Air Quality Update (IAQU), sampling methods typically used for various levels of volatility have been identified. Four categories of volatility can be defined by their boiling point ranges; however, the means by which the compounds are collected, in practice, dictate the categories. The categories and methods of collection are presented in Table A-3, reproduced from IAQU by permission.

Patterns or groupings of chemicals through correspondence analysis (Noma, 1988) of within building and between building plots show significant difference between sick and healthy buildings. In essence, Noma found the difference between a sick building and a control healthy building was not the mean concentration of contaminants, but rather the way the air handling systems controlled the concentrations and the resulting patterns or gradients.

TABLE A-3. CLASSIFICATION OF ORGANIC INDOOR POLLUTANTS[a]

DESCRIPTION	BOILING POINT RANGE[b] FROM °C	TO °C	SAMPLING METHODS TYPICALLY USED IN FIELD STUDIES
Very volatile (gaseous) organic compounds (VVOC)	< 0	50-100	Batch sampling, adsorption on charcoal
Volatile organic compounds (VOC)	50-100	240-260	Adsorption on Tenax, graphitized carbon black or charcoal
Semivolatile organic compounds (SVOC)	240-260	380-400	Adsorption on PUF[c] or XAD-2
Organic compounds associated with particulate matter (particulate organic matter [POM])	> 380		Collection on filters

a Adapted from World Health Organization (1989)

b Polar compounds are at the higher side of the range

c Polyurethane foam

Source: Indoor Air Quality Update newsletter, Arlington, MA.

NOTIFICATION

Under the Right to Know law, employees must be informed about certain VOCs they use, or may be exposed to, in the workplace.

CONTROL AND REMEDIAL PROCEDURES

Many VOCs defy individual treatment. Increased ventilation is frequently used to reduce concentrations. Janssen, chairman of the ASHRAE committee to revise the old ventilation standard, 62-1981, indicated an effort to control these potentially harmful gases prompted a move from 5 cfm/person to 15 and 20 cfm/person.

Selective purchasing of construction materials, furnishings, maintenance and operational materials can avoid or reduce levels of VOC emissions. These materials such as cleaners should be stored in well - ventilated places away from occupied areas.

"Bake-out" procedures (high temperatures to encourage off-gassing followed by air purge) have been found effective in reducing VOCs associated with new construction and refurbishments in some studies. A recent Georgia Institute of Technology study found no appreciable difference.

Time of use can be a key factor. Floor wax, for example, has a very high initial emission factor, which is followed by low - level steady state emissions. EPA research has shown floor wax emissions drop from 10,000 $\mu g/cm^2$ to about 500 $\mu g/cm^2$ in about an hour, and fall below 10 $\mu g/cm^2$ in 10 hours.

Direct exhaust or additional ventilation should be used for activities known to have high VOC emissions, such as spray painting. These activities should be conducted away from occupied zones whenever possible.

REFERENCES

Bayer, C.W., and Black, M.S. "Capillary chromatographic analysis of volatile organic compounds in the indoor environment." J.of Chromatographic Science, Vol 25 (1987), 60-64.

Janssen, John E., "Ventilation for Acceptable Indoor Air Quality," ASHRAE/ JOURNAL, (October 1989), 40-48.

Molhave, L. "Voltile Organic Compounds in Indoor Air Pollutants, " Indoor Air and Human Health, R. Gammage and S. Kaye (eds), Lewish Publishers, Chelsea, MI. (1985), 403-414.

Noma, Elliot; Berglund, Birgitta; Berglund, Ulf; Johansson, Ingegerd; and Baird, John C., "Joint representation of physical locations and volatile organic compounds in indoor air from a healthy and a sick building." Atomospheric Environment Vol. 22, No. 3. (1988), 451-460.

Wallace, L. et. al. "Personal Exposure to Volatile Organic Compounds." Env. Res. Vol. 35 (1984), 193-211.

For further information

Public Works Canada has evaluated the different methods of measuring VOCs and procedures for applying these methods in the field.

Kerr, G. (Public Works Canada, Ottawa, ON (Canada)). 1988. 71p. (CE-02707). Canada. Public Works Canada, Information Research & Library Services, Sir Charles Tupper Bldg., Confederation Heights, Ottawa; ON, CAN K1A OM2; $N/C; MF CANMET/TID, Energy, Mines and Resources Canada K1A OG1; $10 CAN.

The VOC study by Harvard of Kanawha County offers a good case study of measuring outside and indoor VOC levels. To order: Spengler, John D.; Sullivan, Nancy; Ozkaynak, Haluk; Ware, James H.; Cohen, Martin A.; Ryan, P. Barry; "Report on Ambient Exposures to Volatile Organic Compounds in the Kanawha Valley," Harvard University #E-89-06, ($15.00.)

APPENDIX B

SAMPLE CONTAMINANT PROTOCOL *

PAINT

1. Description

Paint provides color and protection for a wide variety of surfaces. Chemists often classify paints according to the way they cure (dry). The two major groups of paint in use today are water-based and solvent-based. Both of these types cure through the evaporation of solvents. All paints have a solid phase known as the pigment, which is responsible for the color, and a liquid phase called the vehicle or binder.

2. Sources:

There are approximately 1,200 companies that manufacture paint in the United States. These companies produce a total of about 990 million gallons of paint annually.

Water-based (latex) paints are not flammable and have little odor. Solvent-based (oil) paints are generally combustible and have a distinctive odor. Latex paint fumes generally have a lower toxicity than solvent-based paints.

3. Health Effects:

There are three primary routes of entry for paint; dermal (skin contact), inhalation (breathing of fumes) and ingestion (eating or swallowing paint). The most common route is inhalation.

Water-based paints, which are generally low in odor, can be used in most areas with minimum risk to an individual's health. However, if an in-

* Source: Anne Arundel County Public Schools (AACPS)

dividual is working in a confined or poorly ventilated area, exposure may cause headache and nausea. When these symptoms occur, the individual should immediately remove themselves from the area and seek fresh air.

Solvent-based paint, which has strong odors, can be used safely by following material safety data sheet information, that is provided by paint manufacturers. When solvent-based paints are used in confined or poorly ventilated areas, an individual can experience irritation of the respiratory tract or acute nervous system depression characterized by: headache, dizziness, staggering, confusion or coma. Inhalation by persons with respiratory problems could have the condition additionally aggravated.

4. Methods of Control:

One effective method of control is the use of a painting protocol which gives proper notification to the school administration, parents and students, and minimizes exposure. (The protocol used by AACPS is shown at the end of this appendix.)

LEAD PAINT

1. Description:

Before 1950, most paints used lead. Lead was used to make several colors, including white, and was used to help paint dry to a hard, durable surface. The use of lead was reduced during the 1960's, but it wasn't until 1977 that Federal Regulations eliminated lead from paints in use for consumers. As lead builds up in the body, it can affect the brain, red blood cells, kidneys and other parts of the body.

2. Sources:

As a coat of paint gets older, it will crumble, chip and peel. Lead from deteriorating paint gets into dust. Lead dust can end up in a child's mouth. Children may even chew on paint chips or painted surfaces. Children may, also, be exposed to paint dust, chips, and fumes from the rehabilitation of older homes, demolition or structural sand blasting of paint work on bridges, elevated water storage tanks and other steel structures. Workers removing old paint can, if they do not protect themselves, breathe in large amounts of lead from dust and fumes.

3. Health Effects:

Lead poisoning in children is a serious problem. Children may not return to normal after treatment or removal from exposure.

At low levels, lead causes long lasting subtle learning, behavioral and psychological problems. Lead also affects the blood forming organs causing anemia and can damage other body systems such as the kidneys and reproductive organs.

Adults are less susceptible to lead poisoning. They are less likely to have irreversible changes in mental status. Symptoms of lead poisoning in adults include loss of appetite, weight loss, insomnia, headache, irritability and abdominal, muscle and joint pains.

Lead poisoning symptoms tend to be more severe during the summer months. In children, the occurrence and severity of symptoms seems to depend upon the exposure. The earliest symptoms are irritability, loss of appetite and decreased play activity.

4. Laws and Regulations

Health and Environmental Article of the Annotated Code of Maryland

6-3-1 Lead Paint Use (effective date 1957)

This law prohibits the use of lead based paint on interior surfaces to which children are exposed and porches of any dwelling.

COMAR (Code of Maryland)

26.02.07 Procedures for Abating Lead Containing Substances from Buildings

This regulation establishes appropriate techniques for abatement of lead paint from interior or exterior areas in residences and group day-care centers.

5. Lead Paint Hazard Determination:

To determine lead paint hazard in a building, painted interior and exterior surfaces should be tested in an environmental survey. The school should be made aware of all lead paint present in the building along with an assessment of hazards.

When a child has been identified as having lead poisoning, the sampling laboratory is required to submit a copy of the lab analysis to the Maryland Department of the Environment. A medical report is required by the Maryland Department of Education. The appropriate health agency will investigate the circumstances. For children who have lead poisoning the sources of lead must be identified and abated. The primary dwelling may not be the source of the hazard, so information regarding prior residence, day care, homes of friends and relatives should be gathered. Secondary sources of lead should also be explored.

6. Measurement:

The two most common types of testing for lead content in paint and by using an X-RF Analyzer (x-ray fluorescence) and by collecting paint scrapings for lead analysis.

The X-RF Analyzer is a field instrument which measures lead content of painted surfaces without removing paint or damaging the surface. The instrument is capable of taking measurements through many layers of paint and giving an immediate on-site reading.

At this time gathering paint scrapings is the most reliable and accurate method for measuring lead content. Wipe sampling can be used to determine the percent of lead in dust on walls and ceilings.

7. Equipment Requirements

Equipment needed for a lead paint abatement project would depend heavily upon the method used for the abatement. Some common tools and equipment needed include 6 mil. plastic for proper containment, respirators (with HEPA filters), disposable clothing, heat guns (Flameless), "Peel Away" (paint removal system), HEPA VAC (special vacuum cleaner for lead dust) and Trisodium Phosphate (TSP) High Phosphate Detergent to clean abated area.

8. Methods of Control:

New Construction - Refer to Health and Environmental Article of the Annotated Code of Maryland Sec. 6-301.
Renovation - Methods to deal with removal of lead paint vary depending on the status of the facility. Control can range from simple encapsulation to total removal depending on the severity of the condition. While there are no specific guidelines, the ideal situation, ignoring cost, is to totally remove the material containing lead paint in those areas of the school where children below the age of 6 can be potentially exposed. Removal of lead paint beyond this criteria would depend on the condition of the facility and cost considerations. Requirements for abatement methods and safety precautions for workers and equipment are contained in Appendix A, Title 26, Department of the Environment, 26.02.07 Procedure for Abating Lead Containing Substance from Buildings.

9. For Further Information:

- AACPS Maintenance Division (301) 255-2535
- AACPS Office of Health Issues (301) 224-5415

PAINT PROTOCOL
ANNE ARUNDEL COUNTY PUBLIC SCHOOLS
MAINTENANCE DIVISION

This protocol, with attachments, is provided for your use in preparation for the interior painting of your school. Based on past experience, certain steps must be taken by the school prior to the beginning of work to ensure that minimum disruption occurs to the teaching process and maximum information is communicated to the parents.

The Paint Foreman or his designated representative will meet with you no later than one month prior to the start of work to coordinate the effort and cover any items which may be of interest to you. As a minimum, he will need your by room painting priority to schedule his work force in your school. Clearly, classrooms or other confined space must be vacated throughout the painting process. The attached Material Safety Data Sheets will provide information on drying time and other precautions which must be taken. Because of the paint we use, overnight drying is generally all that is required for any area being treated. Thus, if a classroom or other confined space is completed in the afternoon, it will be ready for occupancy the following morning. However, the principal reserves the right not to occupy a classroom until he/she is satisfied occupancy will not pose a hazard to the students. The paint crew generally can complete the average classroom in one day.

Your custodial staff will be required to remove all items from storage rooms no later than 7:30 a.m. the day that room is to be painted. Classrooms will not require the removal of furniture. Common areas such as hallways, gymnasiums and media centers will not require the removal of children or furniture. Rather, close coordination between your staff and the paint crew is needed to ensure that children are kept as far from the area being painted as possible during all painting. The paint crew has been instructed to complete a wall or localized area entirely before moving to another part of that common area facility. Thus, if a wall is being painted in one of these areas, children receiving instruction should be at the other side (end) of the space so to be as far removed as possible. Each paint crew assigned to a particular school, will have the use of floor fans, to help ventilate any or all of the paint fumes generated while painting classrooms and/or related areas.

As it pertains to cafetoriums, painting will not start until after the lunch feeding is completed. This means painting will occur only two to three hours in the afternoon each working day. Clearly, the painting of this facility will take longer than in any other part of the school.

The attached notices to parents are provided for school consideration. All parents must be notified by some means. Two choices are offered, one in the form of an official memorandum; the other, less formal, was designed to be a "flyer."

You may have one or mare parents express concern over their child's presence in school during the painting. Since the duration to paint the interior of your school will be a minimum of three months, you and the parents(s) will have to mutually resolve this issue on a case by case basis. The Office of Health Issues, 224-5415, should be contacted for assistance if such concerns do arise.

If you have any questions concerning this memo or the painting process, please contact the Paint Department at 255-2535, or cover your questions or concerns at the scheduled pre-painting meeting.

ANNE ARUNDEL COUNTY PUBLIC SCHOOLS (NOTICE TO PARENTS)

Name
Address
City, State Zip Code

Dear Parents,

The Maintenance Division plans to start painting in our school on or about ______________.

The health and safety of all students is our primary consideration as the painting crews undertake this maintenance and beautification project. The school plant is an important factor in the functioning of the total educational program. Proper maintenance of school buildings is necessary in providing a healthy and pleasant atmosphere. The majority of paint being used in Anne Arundel County Public Schools is latex water-based. Some heavy traffic areas and trim will require the use of oil-based paint. Paints containing lead are never used in school painting.

Instructional areas will be empty during painting and drying times, and children will not be in the cafeteria while it is being painted.

The anticipate that paint crews will be in the school for a period of ______________, beginning on ______________.

If you have any questions or concerns about the scheduled painting, please feel free to contact me at ____________.

With the cooperation of parents, students, and staff, this painting project will result in a bright, new look for our school. I hope you will stop by when the project is completed to see the results for yourself!

Sincerely,

Principal

Source: INDOOR AIR QUALITY MANAGEMENT PROGRAM
ANNE ARUNDEL COUNTY PUBLIC SCHOOLS

APPENDIX C

Investigation Forms

INTERVIEW FORM

PERSON INTERVIEWED ______________________ DATE __/__/__

COMPLAINT ON FILE [ ] YES [ ] NO

BUILDING/AREA WHERE PERSON WORKS ____________________

INTERVIEWER ______________________________________

---

1. How long have you worked in building? __________ In area? ______
   How much time elapsed before symptoms started? ______________

2. Do you have a history of related medical problems (allergies, asthma, respiratory ailments, hayfever, mygrains, eye irritation, eczema, dermatitis, etc.)? If yes, when do they occur?

3. Is any medication taken related to this medical problem?

   [ ] Prescribed [ ] Over the counter

4. Symptoms experienced and estimated duration:

	0-24 Hrs	< 1 wk	1-4 wks	4 wks
Headache				
Eye irritation				
Nose irritation				
Throat irritation; upper respiratory				
Dry mouth				
Backache				
Shortness of breath				
Chest pains				
Nausea				
Flu-like symptoms				
Fever				
Fatigue, malaise lethargy				
Drowsiness				
Dizziness or faintness				
Difficulty concentrating				
Skin dryness, rash, irrita-tion				
Too hot				
Too cold				

Describe symptoms checked above in more detail:

5. Symptom patterns:

Symptoms occur [ ] intermittently [ ] continually

If intermittent,

How often # _____ / (day, week, month) [circle one]

How long do symptoms last (several minutes, several hours, all day, all week)

What months of the year have the symptoms been experienced? (Circle all that apply.)

J F M A M J J A S O N D

What days of the week are they most likely to appear? (Circle all that apply.)

S M T W T F S

What time of the day are they most apt to appear? (Circle all that apply.)

All the time Any time A.M. P.M.

Do symptoms vary in intensity? [ ] Yes [ ] No

If yes, when is problem greatest? ______________________________

Do any of the following apply?

Wears contact lenses	[ ] Yes [ ] No
Operates visual display terminal at least 10% of the day	[ ] Yes [ ] No

Operates copiers at least 10% of the day [ ] Yes [ ] No
Engages in intensive paper handling, especially carbonless sensitive paper [ ] Yes [ ] No
Uses special office equipment [ ] Yes [ ] No
If yes, specify ________________________________

Are symptoms experienced away from work? [ ] Yes [ ] No

Previous work locations? [ ] Yes [ ] No
At home? [ ] Yes [ ] No

6. What are the weather conditions when your symptoms are most apt to appear? Or, are worst?

[ ] Calm, mild [ ] Windy [ ] Cold
[ ] Rainy, stormy [ ] Hot, humid [ ] Dry

7. Are there any specific work activities you engage in just prior to experiencing these symptoms? Are they more apt to happen in a certain work area?

8. How would you describe conditions around your work area? (Check terms, or terms similar to those, used by the person.)

	stuffy		too smokey		too dry
	too drafty		too much glare		too bright
	too humid		too much noise		poor light
	feet too cold		back too cold		back too hot

9. Have you noticed any unpleasant odor(s)? Describe:

10. Is smoking allowed in the work area? [ ] Yes [ ] No

    Do you smoke? [ ] Yes [ ] No

    Are you bothered by smoke? [ ] Yes [ ] No

11. Have you sought medical attention related to the symptoms?
    If yes, describe: [ ] Yes [ ] No

12. What do you think causes your symptoms?

Signatures: Investigator ______________________________ Date __/__/__

Interviewee ______________________________ Date __/__/__
(optional)

PRELIMINARY ASSESSMENT FORM

BUILDING ______________ SPECIFIC AREA(S) ______________

DATE OF INVESTIGATION ______________________________

INVESTIGATOR ______________________________________

A. NATURE AND SCOPE OF COMPLAINTS

(Summary of interview information)

Number interviewed with complaints ______________________

Number interviewed that have no complaints ________________

Symptoms evidenced (indicate number in box):

Frequency by Time Periods

	0-24 Hrs	< 1 wk	1-4 wks	4 wks
Headache				
Eye irritation				
Nose irritation				
Throat irritation; upper respiratory				
Dry mouth				
Backache				
Shortness of breath				

	0-24 Hrs	< 1 wk	1-4 wks	4 wks
Chest pains				
Nausea				
Flu-like symptoms				
Fever				
Fatigue, malaise lethargy				
Drowsiness				
Dizziness or faintness				
Difficulty concentrating				
Skin dryness, rash, irritation				
Too hot				
Too cold				

Describe areas or activities that seem to be associated with increased episodes:

Discuss any distinctive patterns experienced by more than one person for a particular symptom:

Symptom	No. of People	Pattern (Time of day, weather, activity, area)

Is smoking allowed in the area of concern? [ ] Yes [ ] No

Have odors been detected? [ ] Yes [ ] No
Describe odor and indicate number reporting it:

Indicate number who described the work area using any of following terms:

	stuffy		too smokey		too dry
	too drafty		too much glare		too bright
	too humid		too much noise		poor light
	feet too cold		back too cold		back too hot

B. BACKGROUND ASSESSMENT

(A walk through inspection of problem areas and HVAC supply equipment; no measurements.)

OBSERVATIONS: Time: ___:___ A.M.
___:___ P.M.

Temperature in area (thermostat reading is optional)__________ °F

Does it seem [ ] too hot [ ] too cold [ ] drafty?

Humidity (no measurement) [ ] too moist [ ] too dry

Has there been any recent:

	painting		carpet installation		pesticides used
	cleaners used		new construction		new furniture
	wall covering		draperies		other

Is there any evidence of:

	water damage		excessive noise		poor lighting
	mold growth		dirt near ducts		glare

Elaborate on areas checked ______________________________________

____

____

Check basement area or floor on grade for proper drainage and existing or potential leaks. Note problems ______________________________

____

Note any excessive dust/particles ______________________________

____

Describe noticeable odors and suggest possible sources: ___________

____

Check for proper storage and use of cleaning agents, pesticides or special supplies, such as photographic supplies, hazardous materials _________

____

Identify any special equipment that may be a potential pollutant source (copiers, laser printers) ______________________________

____

Describe any episodic or unusual events, such as roof leaks

Date	Event	Comments

5. Describe any previous IAQ problems and investigations.

____

____

6. Describe any actions taken to date to remedy current problem.

____

____

DESIGN, CONSTRUCTION

1. Year built ______________ original sq.ft. ______________

   Site orientation and adjacent land use ______________________________

   ____

   Basic construction ______________________________

Type of windows ____________________ Do they open? _________

Number of stories ____________________ Basement? [ ] Yes [ ] No

Addition(s) _________________________sq.ft.________________

State original design: occupant level ______________________ and functional purposes ______________________________________

If different from original construction, describe structural changes

____

____

Cite any changes in functions/programs _______________________

____

Current occupant level more than planned? [ ] Yes [ ] No

Does this area or a nearby area have:

	research labs		cigarette smoking		copiers
	motor vehicles		animals		garbage
	graphics mat.		other ___________________________		

Describe any other external or attached pollution sources (garages, loading docks, roads, adjacent building's exhaust)

____

2. Modifications and major maintenance:

	Year	Comment
New carpet		
Equipment chgs.		
Furniture		
Wall covering		
Plastics		
Painting		
Bldg. function		
Other ________		

3. Envelope type

Walls ____

Floors ____

Roof, ceiling ____

Does the building have sprayed or foamed insulation? ____________

If yes, when was it applied? _____________________________

HVAC SYSTEM

> **NOTE:** The following HVAC section should be completed by a person, who is knowledgeable about proper operation and maintenance of HVAC systems.

Is building served by one HVAC system? [ ] Yes [ ] No

If portions of the building are served by different units, describe which units serve the areas of concern.

____

____

Heating:

Does facility have an on-site boiler? [ ] Yes [ ] No
If yes, what is approximate age of boiler? ____________________

What is heating fuel source? (Check all that apply.)

[ ] natural gas [ ] coal [ ] electricity [ ] oil #___

Is any auxilliary heat used in area? [ ] Yes [ ] No
If yes, describe ______________________________________

____

What is general condition of the boiler? ____________________

____

Cooling:

Type of cooling used:

[ ] none [ ] central plant [ ] zone [ ] individual unit

If central plant, what are the chilled water temperature settings?

____

Ventilation:

Is there a central ventilation system? [ ] Yes [ ] No

Are windows operable? [ ] Yes [ ] No

Do other areas of the building share the same conditioned air as the area where symptoms have occurred? [ ] Yes [ ] No

If yes, list those areas that you think warrant further investigation

____

____

Indicate type of ventilation system (VAV, dual duct) ________________

____

Where are air intakes? Are they unobstructed? Are they functioning properly? ____

List any outside sources of contamination where emissions may be entering the building ventilation system. ______________________

How does exhaust leave the building? ______________________

Are there processes or activities in the building that may serve as a contaminant source? [ ] Yes [ ] No

If yes, are they vented directly? [ ] Yes [ ] No

What is current outside air setting? ______________________________

Are outside air dampers on air handling units fixed in the closed position? [ ] Yes [ ] No

Or, are units not providing the outside air for another reason? [ ] Yes [ ] No

If yes, what? ______________________________________

Are negative pressure conditions creating infiltration of contaminated air? [ ] Yes [ ] No

Do drain pans have proper inclination for drainage? [ ] Yes [ ] No

Are the linings of the ductwork clean? [ ] Yes [ ] No

Filtration:

Describe type of filtration used, its location in the system, and any scheduled filter maintenance.

____

Are filters clean and accessible? [ ] Yes [ ] No

Humidification/Dehumidification:

Is humidifcation equipment used? [ ] Yes [ ] No

If yes, is it central? ______________________ Or local? __________

Circle type: steam injection air washer water spray

Is dehumidification equipment used? [ ] Yes [ ] No

If yes, is it central? ______________________ Or local? __________

Index